Osama Mohammed Elmardi Suleiman Khayal

Tribology in Vehicles

Osama Mohammed Elmardi Suleiman Khayal

Tribology in Vehicles

Tribological Principles

Noor Publishing

Imprint
Any brand names and product names mentioned in this book are subject to trademark, brand or patent protection and are trademarks or registered trademarks of their respective holders. The use of brand names, product names, common names, trade names, product descriptions etc. even without a particular marking in this work is in no way to be construed to mean that such names may be regarded as unrestricted in respect of trademark and brand protection legislation and could thus be used by anyone.

Cover image: www.ingimage.com

Publisher:
Noor Publishing
is a trademark of
International Book Market Service Ltd., member of OmniScriptum Publishing Group
17 Meldrum Street, Beau Bassin 71504, Mauritius
Printed at: see last page
ISBN: 978-620-0-77981-6

TRIBOLOGY IN VEHICLES

TRIBOLOGICAL PRINCIPLES

Author
Dr. Eng. Osama Mohammed Elmardi Suleiman Khayal

Nile Valley University, Atbara, Sudan
Faculty of Engineering and Technology
Mechanical Engineering Department

June 2020

Dedication

In the name of Allah, the merciful, the compassionate

All praise is due to Allah and blessings and peace is upon his messenger and servant, Mohammed, and upon his family and companions and whoever follows his guidance until the day of resurrection.

To the memory of Professor Sabir Mohammed Salih, Professor Elfadil Adam Abdallah, Associate Professor Mohi-Eldin Idris Habra, Associate Professor Hashim Ahmed Ali, Associate Professor Abdel-Jaleel Yousef, Lecturer Ishraga Salih, Lecturer Intisar Abdu and Associate Professor Salah Ahmed Ali who they taught me the greatest value of hard work and encouraged me in all our endeavors.

This textbook is dedicated mainly to undergraduate engineering students, especially mechanical and production engineering students where most of the applications presented are focused on tribology, which includes the study and application of the principles of friction, lubrication and wear. Tribology is highly interdisciplinary. It draws on many academic fields, including physics, chemistry, materials science, mathematics, biology, and engineering. People who work in the field of tribology are referred to as tribologist.

Last but not least, may Allah accepts this humble work and i hope that it will be beneficial to its readers.

Acknowledgements

I am grateful and deeply indebted to Mechanical Engineer Professor Dr. Mahmoud Yassin Osman for his close supervision, consultation and constructive criticism, without which this work would not have been accomplished.

I am also indebted to many people. Published texts in automotive tribology have been contributed to the author's thinking. The present study has been taken from the point of view of the reciprocating internal combustion engine, transmission and drive line of vehicles, the tires, the brake system, windshield or windscreen wipers, automotive lubricants, and finally the conclusions and the important findings of this study.

Members of Mechanical Engineering Department at Faculty of Engineering and Technology, Nile Valley University - Atbara have served to sharpen and refine the treatment of the topics. The author is extremely grateful to them for constructive criticisms and valuable suggestions.

The author would like to acknowledge with deep thanks and gratitude the moral and financial support extended to them by Nile Valley University via its Faculty of Engineering and Technology despite its financial hardships.

My thanks is extended to Professor Mohammed Ibrahim Shukri for giving his great experience in writing reports, researches and books according to the standard format, and also for revising several times the manuscript of the present book.

Special gratitude is due to Professor Mahmoud Yassin Osman

for continuous follow up step by step the completion of this textbook and the greatest advice he has given in writing sequentially the various chapters of this book.

I express my profound gratitude to Mr. Osama Mahmoud Mohammed Ali of Dania Center for Computer and Printing Services, Atbara, and engineer Awad Ali Bakri who they spent many hours in editing, re – editing and correcting the manuscript.

Finally i would like to acknowledge the generous assistance of the staff at different levels working in the industrial liaison and consultancy unit, Atbara and the mechanical and production engineering departments' staff working in the different workshops and laboratories in the Faculty of Engineering and Technology, Atbara.

Contents

Preface

The vehicle is one of the most common machines in use today, and it is no exaggeration to state that it is crucial to the economic success of all the developing and developed nations of the world and to the quality of life of their citizens. The vehicle itself consists of thousands of component parts, many of which rely on the interaction of their surfaces to function. There are many hundreds of tribological components, from bearings, pistons, transmissions, clutches, gears, to wiper blades, tires, and electrical contacts. The application of tribological principles is essential for the reliability of the vehicle, and mass production of the vehicle has led to enormous advances in the field of tribology. For example, many of the developments in lubrication and bearing surface technology have been driven by requirements for increased capacity and durability in the vehicle industry. For the purpose of classifying the tribological components, one can divide the motor vehicle into engine, transmission, drive line, and ancillaries such as tires, brakes, and windshield wipers. In the following chapters, each automotive component is discussed in detail. Lubricants used with these automotive tribological components are described in the last chapter of this study.

The aim of the present textbook was to review the importance of automotive tribology from the point of view of the reciprocating internal combustion engine, transmission and drive line of vehicles, the tires, the brake system, windshield or windscreen wipers, automotive lubricants, and finally the conclusions and the important findings of this study.

Chapter one constitutes general introduction to the engine of vehicles. This chapter includes introduction, importance of engine tribology, lubrication regimes in the engine, engine bearings, piston assembly, valve train, and future developments.

In chapter two transmission and drive line of vehicles were introduced and discussed thoroughly from the consideration of transmission, traction drive, universal and constant velocity joints, wheel bearings, and drive chains.

Chapters three discusses the tires of automotive from the viewpoint of introduction, the effect of introduction, construction of a tire, mechanics of load transfer, contact area and normal pressure distribution, slip and generation of shear traction, hydroplaning, and wear characteristics.

Chapter four discusses the braking system of vehicles which includes a brief introduction to braking system, contact pressure and shear traction, materials, friction, and wear.

Chapter five discusses windshield or windscreen wipers. The primary function of the wiper is to improve visibility by removing the dirt particles on the windshield by a wiping action. This is usually accomplished with the windshield washer on. The chemicals in the fluid help in loosening dirt particles, which are then scraped by the blade.

The conclusions which is the sixth and last chapter of this research study summarizes the important points and findings in automotive tribology which were reached in this textbook.

Keywords: Engine tribology, transmission, tire, brakes, lubricants.

Chapter one
The Engine of Vehicles

1.1 Introduction

Tribology is the science and engineering of interacting surfaces in relative motion. It includes the study and application of the principles of friction, lubrication, and wear.

Tribology is highly interdisciplinary. It draws on many academic fields, which includes different subjects such as physics, chemistry, materials science, mathematics, biology, and engineering. People who work in the field of tribology are referred to as tribologist (Jost, Peter, 1966).

Despite the relatively recent naming of the field of tribology, quantitative studies of friction can be traced as far back as 1493, when Leonardo da Vinci first noted the two fundamental 'laws' of friction (Hutchings, Ian M., 2016). According to da Vinci, frictional resistance was the same for two different objects of the same weight but making contact over different widths and lengths. He also observed that the force needed to overcome friction doubles as weight doubles. However, da Vinci's findings remained unpublished in his notebooks (Hutchings, Ian M., 2016).

The two fundamental 'laws' of friction were first published (in 1699) by Guillaume Amontons, with whose name they are now usually associated. They state that (Hutchings, Ian M., 2016): the force of friction acting between two sliding surfaces is proportional to the load pressing the surfaces together, and the force of friction is independent of the apparent area of contact between the two surfaces.

Although not universally applicable, these simple statements hold for a surprisingly wide range of systems (Gao, Jianping; Luedtke, W. D.; Gourdon, D.; Ruths, M.; Israelachvili, J. N.; Landman, Uzi, 2004). These laws were further developed by Charles-Augustin de Coulomb (in 1785), who noticed that static friction force may depend on the contact time and sliding (kinetic) friction may depend on sliding velocity, normal force and contact area (Dowson, Duncan, 1997) and (Popova, Elena; Popov, Valentin L. ,2015).

In 1798, Charles Hatchett and Henry Cavendish carried out the first reliable test on frictional wear. In a study commissioned by the Privy Council of the UK, they used a simple reciprocating machine to evaluate the wear rate of gold coins. They found that coins with grit between them wore at a faster rate compared to self-mated coins (Chaston, J.C., 1974). In 1860, Theodor Reye (Rühlmann, Moritz, 1885) proposed Reye's hypothesis (Reye, Karl Theodor (1860). In 1953, John Frederick Archard developed the Archard equation which describes sliding wear and is based on the theory of asperity contact (Archard, John Frederick, 1953).

Other pioneers of tribology research are Australian physicist Frank Philip Bowden (Tabor, D., 1969) and British physicist David Tabor (Field, J., 2008), both of the Cavendish Laboratory at Cambridge University. Together they wrote the seminal textbook The Friction and Lubrication of Solids (Bowden, Frank Philip; Tabor, David, 2001) (Part I originally published in 1950 and Part II in 1964). Michael J. Neale was another leader in the field during the mid-to-late 1900s. He specialized in solving problems in machine design by applying his knowledge of tribology. Neale was respected as an educator with a gift for integrating theoretical work with his own practical experience to produce easy-to-understand design guides. The Tribology Handbook (Neale, Michael J., 1995), which he first edited in 1973 and updated in 1995, is still used around the world and forms the basis of numerous training courses for engineering designers. Duncan Dowson surveyed the history of tribology in his 1997 book History of Tribology (2nd edition) (Dowson, Duncan, 1997). This covers developments from prehistory, through early civilizations (Mesopotamia, ancient Egypt) and highlights the key developments up to the end of the twentieth century.

The reciprocating internal combustion engine as shown in Figure 1.1 is the prime mover in the motor vehicle, as well as in many other modes of ground and sea transport, including motorcycles, scooters, mopeds, vans, trucks, buses, agricultural vehicles, construction vehicles, trains, boats, and ships. Further applications can be found in the field of electrical power generation, where the internal combustion engine is used for primary and standby electricity generation and for combined heat and power plants.

The popularity of the reciprocating internal combustion engine is testament to its performance, reliability, and versatility. However, there are also some major drawbacks. Thermal and mechanical efficiencies are relatively low, with much of the energy of the fuel dissipated as heat loss and friction. The internal combustion engine is also a significant contributor to atmospheric pollution through hydrocarbon, particulate, and NO_x (nitrogen oxides) emissions and to the greenhouse effect via carbon dioxide (CO_2) emissions. A viable alternative with the required portfolio of attributes including cost, however, has yet to be found and, hence, the reciprocating internal combustion engine is set to dominate the road vehicle market for the foreseeable future.

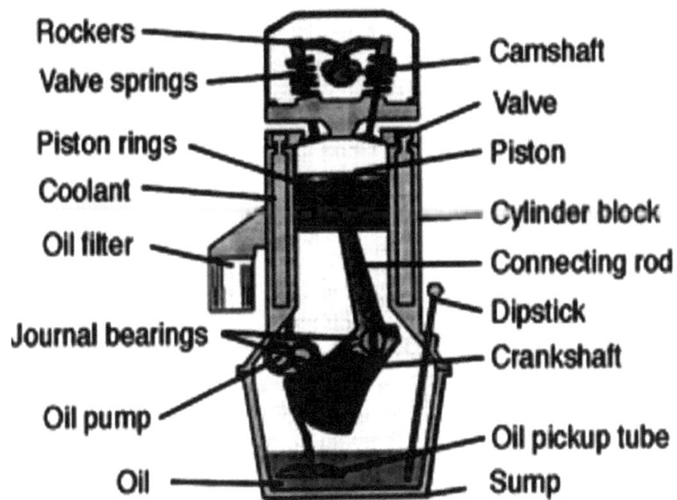

Figure 1.1 Main Engine Components in a Reciprocating Internal Combustion Engine

1.2 Importance of Engine Tribology

To reduce friction and wear, the engine tribologist is required to achieve effective lubrication of all moving engine components, with minimum adverse impact on the environment. This task is particularly difficult given the wide range of operating conditions of load, speed, temperature, and chemical reactivity experienced in an engine. Improvements in the tribological performance of engines can yield:

• Reduced fuel consumption

• Increased engine power output

• Reduced oil consumption

• A reduction in harmful exhaust emissions

• Improved durability, reliability, and engine life

• Reduced maintenance requirements and longer service intervals

With such large numbers of reciprocating internal combustion engines (Figure 1.1) in service, even the smallest improvements in engine efficiency, emission levels, and durability can have a major effect on the world economy and the environment in the medium to long term (Taylor, 1998). It is interesting to consider where the energy derived from combustion of the fuel is apportioned in an engine. In a published paper, (Andersson, 1991) showed the distribution of fuel energy for a medium size passenger car during an urban cycle. Only 12% of the available energy in the fuel is available to drive the wheels, with some 15% being dissipated as mechanical, mainly frictional, losses. Based on the fuel consumption data in Anderson's publication, a 10% reduction in mechanical losses would lead to a 1.5% reduction in fuel consumption. The worldwide economic implications of this are startling in both resource and financial terms and the prospect for significant improvement in efficiency by modest reductions in friction is clear (Taylor, 1998). Concerning energy consumption within the engine as shown in Figure 1.2, friction loss is the major portion (48%) of the energy consumption developed in an engine (Nakasa, 1995). The other portions are the acceleration resistance (35%) and the cruising resistance (17%). If one looks into the entire friction loss portion, engine friction loss is 41% and the transmission and gears are approximately 7%. Concerning engine friction loss only, sliding of the piston rings and piston skirt against the cylinder wall is undoubtedly the largest contribution to friction in the engine. Frictional losses arising from the rotating engine bearings (notably the crankshaft and camshaft journal bearings) are the next most significant, followed by the valve train (principally at the cam and follower interface), and the auxiliaries such as the oil pump, water pump, and alternator (Monaghan, 1987, 1989). The relative proportions of these losses, and their total, vary with engine type, component design, operating conditions, choice of engine lubricant, and the service history of the vehicle (i.e., worn condition of the components). Auxiliaries should not be overlooked, as they can account for 20% or more of the mechanical friction losses. For example, an oil pump in a modern 1600 cc gasoline engine can absorb 2 to 3 kW of power at full engine speed, while for a racing car this can rise to some 20 kW (Taylor, 1998).

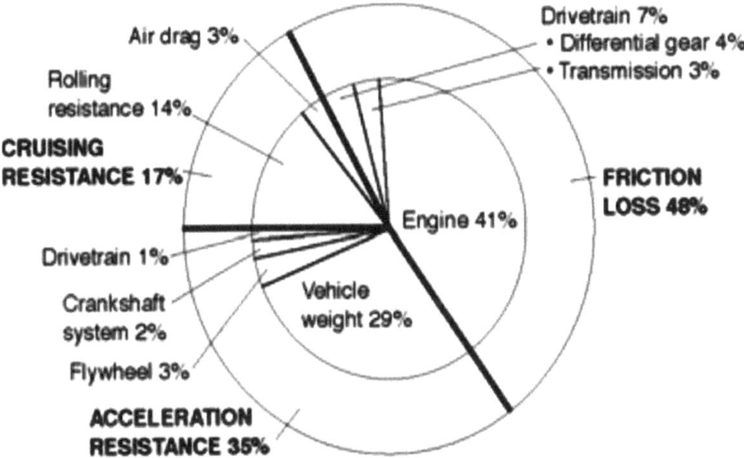

Figure 1.2 Energy Consumption Developed in an Engine

1.3 Lubrication Regimes in the Engine

As with any other system designed to operate with a liquid lubricant, the key operating tribological parameter in an engine is the lubricant film thickness separating the interacting component surfaces. More precisely, it is the relative magnitude of the lubricant film compared to the combined surface roughness of the two surfaces, the so-called film thickness ratio, or parameter, λ :

$$\lambda = \frac{h}{\left(\sigma^2_{surface\ 1} + \sigma^2_{surface\ 2}\right)^{1/2}}$$

Where h is the film thickness calculated through the application of classical thin-film analysis taking the surfaces to be smooth, and σ is the root mean square surface roughness. A related version of this equation uses the center line average surface roughness values, ($Ra_{surface\ 1}$ + $Ra_{surface\ 2}$), in the denominator in place of the root mean square term. Figure 1.3 shows a plot of the relationship between the coefficient of friction and the oil-film thickness ratio. The diagrams at the top of the figure provide a visual example of the lubrication of two surfaces that are in relative motion to each other and that are separated by an oil film. At the box on the left side, there is surface contact; at the box on the right, a fluid film separates the surfaces; and between these two extremes partial, or intermittent, contact occurs. The various lubrication regimes are listed below the diagrams. The boundaries between these regimes are not sharply defined. The curved line below the lubrication regimes indicates the relationship between the friction coefficient and the oil-film thickness ratio. Examples of the lubrication regimes for several automotive components are shown at the bottom of Figure 1.3. Different automotive components rely on different modes of lubrication to achieve acceptable performance, and each may experience more than one regime of lubrication during a single cycle. Generally, journal and thrust bearings are designed to operate in the hydrodynamic lubrication regime in which bearing surfaces are separated by a lubricant film. Actual metal-to-metal contact is expected to take place only at low speeds and high loads and with low-viscosity lubricants. In contrast, valve train, piston ring assembly, and transmission clutch sliding generally take place under mixed or boundary lubrication conditions; surface contact occurs, and chemical films or reaction products may be an important means of surface protection. Furthermore, the importance of different lubrication regimes for each component may change with flattening of the surface roughness in the interface, wear of critical interacting surfaces, and degradation of the lubricant with time.

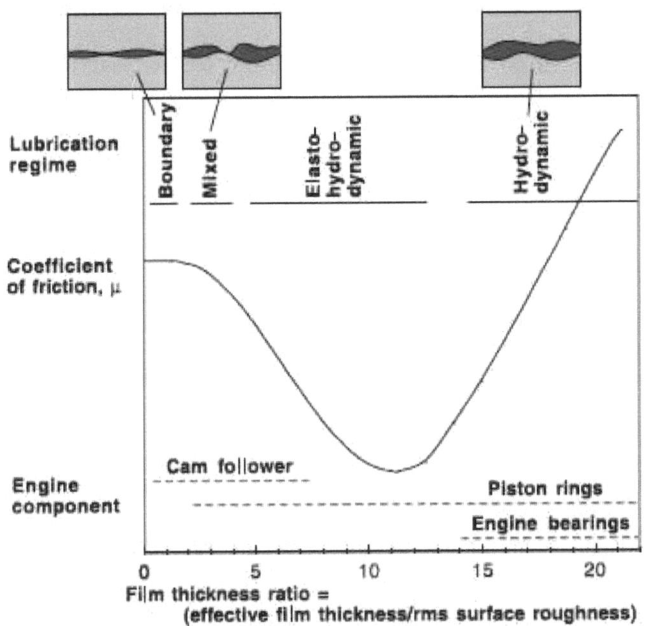

Figure 1.3 Lubrication Regimes for Engine Components and Typical Friction Relationships

1.4 Engine Bearings

Within the engine, rotating journal bearings are used to support the camshaft, the crankshaft, and the connecting rod. The basic construction is a split half-shell bearing fitted into a bore and incorporating some form of locating notch. The various forms the bearing shells may take are summarized in Heizler (1999). The shells are formed from a steel backing strip to give strength and tight dimensional tolerance and overlaid with a relatively soft bearing material such as tin-aluminum or lead-bronze. A further thin soft overlay is then put onto the bearing material to ensure a degree of initial conformability (Massey et al., 1991). Shafts designed for running against engine bearings are generally made of heat-treated steels or spheroidal graphite irons, with a hardness approximately 3 times that of the principal bearing material. Lubricant is directed to the bearings either by jet impingement or by passing through the bore of a hollow shaft. Provided the bearings are adequately lubricated, wear (after an initial running in period) is low. However, shaft misalignment or particulate contamination of the lubricant supply can lead to excessive wear. Bearing corrosion is an additional failure mechanism. As with most components in the engine, the tribology of journal bearings is complicated by issues such as lubricant supply, thermal effects, dynamic loading, and elasticity of the bounding solids. The Mobility technique for the analysis of dynamically loaded engine bearings was established some 30 years ago and remains the most common approach (Booker, 1965, 1969). The technique is robust and has proved amenable to simple computer analysis. It yields, among other results, the cyclic minimum film thickness between the journal and the bearing, which is an important design parameter. However, it is important to note that many simplifying assumptions are implicit in the use of the mobility method, which means that the predictions can only be used as a benchmark. Alternative approaches and research studies in which some of the assumptions have been relaxed are discussed in Martin (1983), Taylor (1998), and Xu (1999). The mobility method and many other alternative methods of solution assume that engine bearings operate entirely in the hydrodynamic lubrication regime. The benchmark prediction for a satisfactory minimum lubricant film thickness in an engine bearing used to be approximately 2.5 μ m. Today, minimum film thickness predictions in the range 0.5 to 1.0 μ m are being made for engine bearings in passenger cars, suggesting that asperity interaction may occur between the journal and bearing for at least part of the engine cycle. The implications of operation in the mixed lubrication regime for the analysis and design of engine bearings are significant and necessitate a fundamentally different approach (Priest et al., 1998; Xu, 1999).

1.5 Piston Assembly

The piston is at the heart of the reciprocating internal combustion engine, forming a vital link in transforming the energy generated by combustion of the fuel and air mixture into useful kinetic energy. The piston carries the ring pack, which is essentially a series of metallic rings, the primary role of which is to maintain an effective gas seal between the combustion chamber and the crankcase. The rings of the piston ring pack, which in effect form a labyrinth seal, achieve this by closely conforming to their grooves in the piston and to the cylinder wall. Secondary roles of the piston ring pack are to transfer heat from the piston into the cylinder wall, and then into the coolant, and to limit the amount of oil that is transported from the crankcase to the combustion chamber. This flow path is probably the largest contributor to engine oil consumption and leads to an increase in harmful exhaust emissions as the oil mixes and reacts with the other contents of the combustion chamber. The desire to extend service intervals of engines and minimize harmful exhaust emissions to meet ever more stringent legislative requirements means that the permissible oil consumption levels of modern engines are very low compared to their predecessors of 10 or 20 years ago (Munro, 1990). The left side of Figure 1.4 is a schematic representation of a typical piston assembly from a modern automotive engine. From a tribological perspective, the main piston features of interest are the grooves, which carry the piston rings, and the region of the piston below the ring pack (the piston skirt), which transmits the transverse loads on the piston to the cylinder wall. The top two piston rings are referred to as the compression rings. Firing pressure pushes these rings out until the entire ring face engages the cylinder wall. Gas pressure is utilized to supplement the inherent elasticity of the ring to maintain an effective combustion chamber seal. The top compression ring is the primary gas seal and, as the ring nearest the combustion chamber, encounters the highest loads and temperatures. The top compression ring usually has a barrel-faced profile with a wear-resistant coating such as chromium or flame-sprayed molybdenum on the periphery and occasionally on the flanks. The second compression ring, which is sometimes referred to as the scraper ring, is designed to assist in limiting upward oil flow in addition to providing a secondary gas seal. The right side of Figure 1.4 illustrates how the second compression ring scrapes surplus oil from cylinder walls and how the ring rides on a film of oil and presents a lower edge to cylinder wall. As such, the second compression ring has a taper-faced, downward scraping profile that is not normally coated. The piston rings at the right of Figure 1.4 are notched. A more conventional design is a ring that is not notched, as in the drawing in the lower left portion of Figure 1.4.

The bottom ring in the pack is the oil-control ring, which has two running faces (or lands), and a spring element to enhance radial load. As its name suggests, the role of this ring is to limit the amount of oil transported from the crankcase to the combustion chamber, and by design it has no gas sealing ability. The periphery of the lands and, occasionally, the flanks are often chromium plated. There are a large number of subtle variations to the basic ring design, a good summary of which can be found in Neale (1994). Notable examples are the keystone compression ring with tapered sides designed to prevent ring sticking due to deposit formation in the piston ring grooves of diesel engines; the internally stepped compression ring, which imposes a dishing on the ring when fitted to improve oil control; and the multipiece, steel rail, oil-control ring. In contrast to the one-piece oil-control ring, the multi-piece oil-control ring has smaller land heights, which increases conformability to the cylinder wall, and hence oil control, with reduced spring force and a ring of reduced overall axial height (Brauers, 1988).

The piston ring is perhaps the most complicated tribological component in the internal combustion engine to analyze. It is subjected to large, rapid variations of load, speed, temperature, and lubricant availability. In one single stroke of the piston, the piston ring interface with the cylinder wall may experience boundary, mixed, and full fluid film lubrication (Ruddy et al., 1982). Elastohydrodynamic lubrication of piston rings is also possible in both gasoline and diesel engines on the highly loaded expansion stroke after firing (Rycroft et al., 1997).

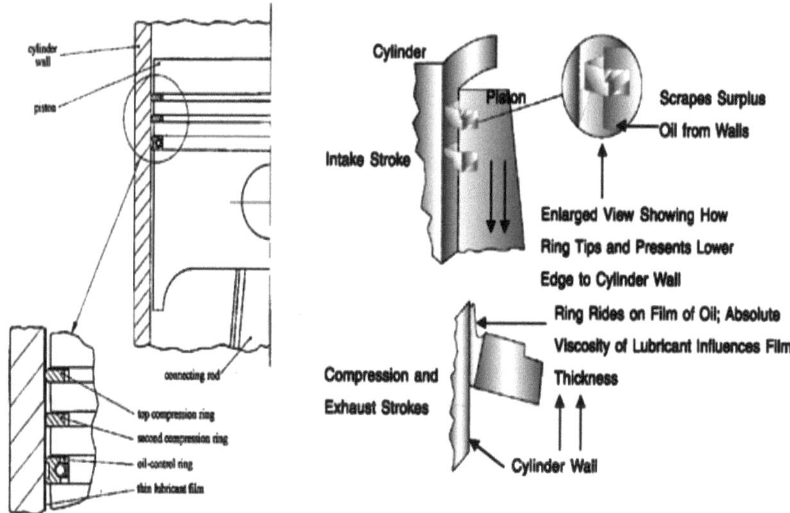

Figure 1.4 Typical Piston Assembly and Piston Ring Function from a Modern Automotive Engine

A typical prediction from a mathematical model such as Ruddy et al. (1982) for the cyclic variation of minimum film thickness between a piston ring and cylinder wall is shown in Figure 1.5, where zero degrees of crank angle is top dead center firing. The prediction is for the top compression ring of a four-stroke gasoline engine assuming a plentiful supply of lubricant. The combined surface roughness of the piston ring and cylinder wall in this case was approximately 0.1 μ m. The solution exhibits a characteristic shape of curve with small film thickness around the dead center positions where the sliding velocity, and hence lubricant entrainment velocity, is low and large film thickness at the mid-stroke positions where the sliding and entrainment velocities are large. Often piston rings are extremely starved of oil, leading to a reduction in the film thickness and a much more complex situation to analyze and interpret (Priest et al., 1999). During the engine cycle, the piston itself exhibits a complex secondary motion, transverse movement toward the cylinder wall, and tilting about the main piston pin axis (Li et al., 1982). This generally results in fluid film or mixed lubrication between the piston skirt and the cylinder wall. Modern, lightweight pistons have skirts that may deflect elastically during interaction with the cylinder wall, leading to the application of Elastohydrodynamic lubrication theory to these components (Oh et al., 1987). Frictional losses between the skirt and the cylinder wall are significant, accounting typically for about 30% of total piston assembly friction, and the interaction with the cylinder wall can lead to noise generation, so-called piston slap. Gray cast iron, carbide/malleable iron, and malleable/nodular iron are the most common base materials for all types of compression piston rings and single piece oil-control rings. However, steel is growing in importance as a piston ring material because of its high strength and fatigue properties and its manufacturing route, which enables rings of small axial height to be produced by, for example, forming steel wire. Steel is used for top compression rings and the rails of multi-piece oil-control rings. There are a multitude of piston ring running-in surface treatments and coatings, although many of these are subtle variations of similar processes. Chromium plating and flame-sprayed molybdenum are the most common wear-resistant coatings, although plasma-sprayed molybdenum, chromium, metal composites, cermets, and ceramics are growing in popularity as their technology progresses. Cylinder walls are generally manufactured from gray cast iron, either plain or with the addition of alloying elements, or an aluminum alloy. The surface treatment and coating of cylinder walls is less common than with piston rings, although proprietary wear-resistant coatings such as Nikasil plating are applied to aluminum bores. Aluminum-silicon alloy is the main material used for automotive pistons, occasionally with additions of other elements to enhance particular properties (e.g., copper to increase fatigue strength). Piston skirt coatings, based on materials such as polytetrafluoroethylene (PTFE) or graphite, are occasionally applied. In contrast to piston ring and cylinder liner coatings, piston skirt coatings are primarily intended to reduce friction rather than wear. Understanding and controlling the wear of the piston assembly is crucial to successful engine performance. Manufacturers through long experience have come to rely on early life wear of the piston rings and cylinder wall to modify the profile and roughness of the interacting surfaces to achieve acceptable performance as part of the running-in process. However, a clear understanding of the complex interactions between lubrication and wear of these components has only recently started to emerge (Priest et al., 1999). In addition to wear as a consequence of mechanical interactions, corrosive wear can occur in the upper cylinder region during short-trip service in a winter climate (Schwartz et al., 1994). The as-manufactured surface finish of piston rings and cylinders can have a major influence on wear behavior and thereby the success or failure of an engine. Although a wide range of surface finishing techniques has evolved for both piston rings and cylinder liners, the objectives of these processes appear to be the same: to improve oil retention at or within the surface, to minimize scuffing, and to promote ring profile formation during the running-in period. Plain cast iron piston rings are often used in a fine turned condition and electroplated; flame-sprayed and plasma-sprayed rings are generally ground to the desired finish. The poor wettability by oil of chromium-plated piston rings is overcome by etching a network of cracks and pores into the surface by reversing the plating current, chemical etching, grit blasting, lapping, or depositing the plate over a surface with fine-turned circumferential grooves. Cylinders are honed with ceramic stones, diamond hones, rubber tools, cork tools, or composite stones of cork or diamond. Piston skirts with a turned finish have been shown to give superior fuel economy and reduced risk of scuffing.

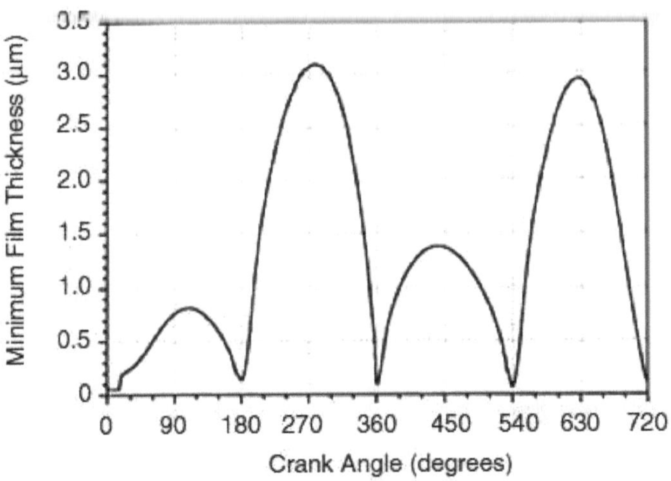

Figure 1.5 Typical Predicted Variation of Film Thickness between a Top Piston Ring and Cylinder Wall of a Gasoline Engine with a Plentiful Supply of Lubricant

1.6 Valve Train

A modern valve train system as shown in Figure 1.6 should include valves, valve springs, valve spring retainers, valve keys, rocker arms, piston rods, lifter/tappets, and a camshaft. The primary function of the valve train system is to transform rotary camshaft motion into linear valve motion in order to control fluid flow into and out of the combustion chamber. The second function is to drive ancillary devices such as distributors, fuel pumps, water pumps, and power steering pumps. Many different styles of valve train mechanisms have been used on engines. Three basic valve train systems are (1) poppet valve, (2) sleeve valve, and (3) rotary valve. In a valve train system, the poppet valve configuration is the most popular and is used by virtually all the major automobile manufacturers for the inlet and exhaust valves of reciprocating internal combustion engines. The opening and closing of the valves is invariably controlled by a camshaft driven from the crankshaft to ensure synchronization of the valve motion with the combustion cycle and piston movement. Some of the typical mechanisms and some of the design variations have been executed on poppet valve systems. Common mechanisms are generally classified and referred to throughout the automotive industry as Type 1 through Type 5 (Nunney, 1998). The basic mechanisms for Type 1 through Type 5, as shown in Figure 1.7, include the following:

1. Direct acting valve train

2. End pivot valve train

3. Center pivot

4. Center pivot with cam follower

5. Overhead (OHV) pushrod

These types can be designed using either roller or sliding friction at each contact interface. There is, however, a large range of mechanisms in use to transmit the motion of the cam to the valve. These include push-rod operated, center-pivoted rocker, finger follower, direct acting bucket follower, and roller follower (Heizler, 1999). Other designs, such as rotary or sleeve valves, have proved difficult to lubricate, cause excessive oil consumption, have poor sealing properties, and generate excessive friction (Buuck, 1982).

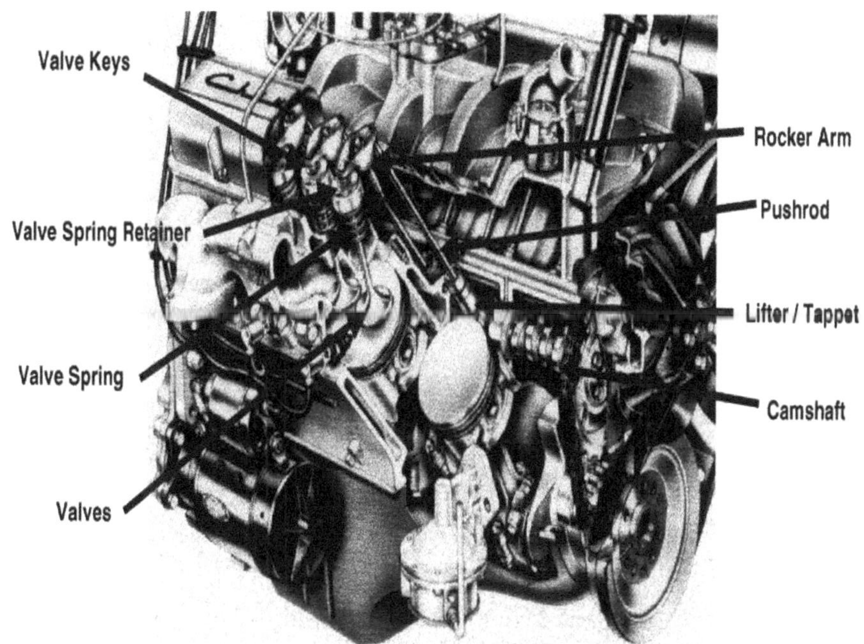

Figure 1.6 Modern Valve Train System

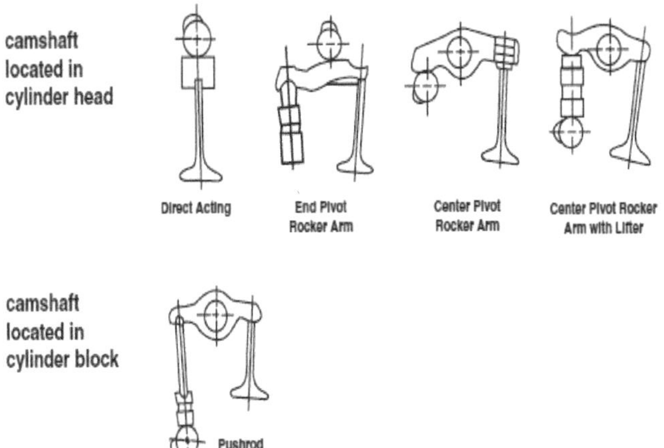

Figure 1.7 Valve Train Mechanisms

A comparison of friction, effective mass or valve weight, maximum engine speed, and overall engine package among the different types of valve train systems is shown in Figure 1.8.

Mechanism Feature	Type 1	Type 2	Type 3	Type 4	Type 5
Natural Frequency (Hz)	2000 -3000	1200 -1500	900 -1400	900 - 1400	400 - 700
Effective Mass @ Valve (gr.)	140 - 160	80 - 120	120 - 160	130 - 170	240 -290
Maximum RPM	6500 ++	6500 +	6000 +	6000 +	4000 - 6000
Friction (A-E)	E	A	B	C -D	C - D
Overall Engine Packaging (A-E)	D - E	D - E	B	C	A

***A= Best , E= Worst**

Figure 1.8 Comparison of Friction, Effective Mass, Engine Speed, and Overall Engine Package among These Five Types of Valve Train Systems

Friction also plays an important role in valve train type selection. The best configuration is roller follower, which can significantly reduce valve train friction. This is because the coefficient of friction at the camshaft/follower interface is less by an order of magnitude for rolling compared to sliding friction. The direct-acting system has poor frictional characteristics because it is not equipped with roller followers. The next most important characteristics after rollers or friction control are a function of the lowest mass, least number of components, and the lowest number of friction interfaces. The friction associated with the valve train is generally considered to be a small component of the mechanical losses, typically 10% compared to the 50% linked to the piston assemblies. However, at low engine speeds and for larger, slower running engines associated with the prestige car market, valve train friction is seen to rise significantly to 20 to 25%. This has been one of the main drivers behind introducing into the engine oil such friction modifier additives as molybdenum dithiocarbamate (MoDTC), which endeavor to reduce friction in the boundary lubrication regime (Korcek et al., 1999). The valve train presents a broad spectrum of problems to the tribologist in relation to the cam and follower, valve guides, valve stem seals, valve seats, hydraulic lash adjusters, lifter guides, pivots, camshaft bearings, belt drives, and chain drives. That said, the most critical interface in the valve train is that between the cam and follower, a contact that has proved difficult to lubricate effectively with all designs of valve train. Traditional design philosophy assumed that the cam and follower operated entirely in the boundary lubrication regime. Hence, attention tended to be limited to the materials and surface treatments of the components and the lubricant additives intended to produce wear-resistant surface films by chemical reaction (e.g., zinc dialkyldithiophosphates [ZDDP or ZDTP]). Scientific studies over the last 20 years or so, however, have shown that mixed and Elastohydrodynamic lubrication have a significant role to play in the tribological performance of cams and followers (Taylor, 1994). These film thickness predictions can be usefully compared with typical surface roughness values for automotive metallic cams and followers of about 0.2 μ m Ra. Although this analysis is relatively simple and neglects potentially important physical effects, it highlights the important message that the cam and follower may experience boundary, mixed, and Elastohydrodynamic lubrication in a single cycle. Substantial film thickness tends to occur on the rising and falling flanks; whereas around the nose, the film thickness tends to be much smaller, with the potential for surface contact and wear. The most common materials for cams and followers are irons and steels with a variety of metallurgies and surface treatments to assist running-in and prevent early life failure. Ceramic followers are, however, becoming more common, especially in direct-acting valve trains, in an attempt to reduce frictional losses (Gangopadhyay et al., 1999). Wear has been a persistent problem with cams and followers for many years, especially with finger follower configurations. Recent advances in our understanding of the link between kinematics, lubrication, and wear of valve trains are now helping to overcome these difficulties (Bell, 1998), as is the trend toward direct acting and roller follower systems. The failure modes of cams and followers are pitting, polishing, and scuffing, all of which are influenced by materials, lubrication, design, and operating conditions. The durability and type of failure can vary considerably, depending on the combination of materials chosen, their surface treatment, and the lubricant and its additive package. Wear is also a problem at the valve/seat interface; the valve "recesses" into the seat and results in loss of engine timing. Loading is from a combination of the dynamic closing of the valve under cam action and the application of the firing pressure on the closed valve face, and therefore the seat is subjected to both high static and dynamic stresses. Temperatures in the region of the valve are high (typically 300°C to 500°C), and lubrication tends to be from small quantities of oil that flow past the valve guide. The valve may be allowed to rotate (usually slowly at about 1 to 2 revolutions per minute) under the action of the cam on the bucket. To reduce wear, an insert of hardened steel is shrunk-fit into the head. Valve and seat materials need to have high strength, wear resistance, temperature stability, and corrosion resistance (Narasimhan et al., 1985). Commonly, inlet valves are made from hardened, low-alloy martensitic steel for good wear resistance and strength. The exhaust valves are subjected to higher temperatures and are often made from precipitation-hardened, austenitic stainless steel for corrosion resistance and hot hardness. The valve seats are made from cast or sintered high-carbon steel. In some engines, the seat is formed by the induction hardening of the cylinder head material.

1.7 Future Developments

The petroleum and automotive industries are facing tough international competition, government regulations, and rapid technological changes. Ever-increasing government regulations require improved fuel economy and lower emissions from the automotive fuel and lubricant systems. Higher energy-conserving engine oils and better fuel-efficient vehicles will become increasingly important in the face of both the saving of natural resources and the lowering of engine friction (Hsu, 1995). The United States Department of Energy conducted a workshop (Fessler, 1999) in which the focus was on industry research needs for reducing friction and wear in transportation. Reducing friction and wear in engine and drive trains could save the U.S. economy as much as $120 billion per year. Recommendations from that workshop (although each sector of the ground transportation industry has different needs and objectives) and desirable areas for future research (including current technical status, goals, and barriers related to fuel efficiency, emissions, durability, and profitability of powertrain systems) are as follows:

1. Develop a quantitative understanding of failure mechanisms such as wear, scuffing, and fatigue. This is important both for developing improved computational design codes and for developing bench tests to predict accurately and reliably the tribological behavior of full-scale automotive components.

2. Develop a variety of affordable surface modification technologies (Tung et al., 1995b) that are suitable for various vehicle components used with different fuels or lubricants under a variety of operating conditions.

3. Develop a better understanding of the chemistry of lubricants and how additives affect the interactions between lubricants and rubbing surfaces. This will provide a foundation for developing new lubricants that will be longer-lasting, environmentally friendly, and capable of handling increased soot and acid loading from EGR (exhaust gas recirculation), compatible with catalysts, and compatible with new, lightweight nonferrous materials.

Stringent federal legislation calling for better fuel economy and reduced emissions is the driving force for improved fuel efficiency and development of new engine technology. Among the various approaches for improving fuel efficiency, the use of energy-conserving engine oils is the most economical way in the automotive industry to acquire the necessary gains, compared to complicated hardware changes (Tung et al., 1995a). In addition, design and legislative pressures for cleaner, more efficient engines with higher specific power outputs is forcing tribological engine components to be operated with generally thinner oil films. One notable trend is the move toward lower viscosity engine oils (e.g., SAE [Society of Automotive Engineers] 5W-20 and 0W-20) in an effort to improve fuel economy. While this helps to reduce friction losses, it also leads to potential durability problems and a more critical role for the surface topography of engine components (Priest et al., 1998). In addition, there is a drive to extend engine service intervals. The engine has therefore to withstand an increasingly contaminated and degraded lubricant. The ability to incorporate more and more aspects of the physical behavior of lubricants into analytical modeling is an important and a fast-developing field (Coy, 1997). Uppermost in this regard are the reduction in viscosity at high shear rate, particularly with polymer-containing multigrade lubricants; the rise in viscosity at elevated pressure; and the boundary friction and wear behavior in the mixed and boundary lubrication regimes. Models of the rate of degradation of oil additives are also becoming available. One of the biggest challenges facing the engine tribology community is to make an effective link between the physical tribology of the components (Tung et al., 1996) and the complex chemical behavior and degradation of engine lubricants, both in the bulk fluid (Bush et al., 1991) and at the surface (Korcek et al., 1999). The engine oils are expected to last longer and simultaneously reduce engine losses.

Chapter two
Transmission and Drive Line of Vehicles

2.1 Transmission

The tribological characteristics of the transmission clutch and band are highly important because they control transmission shift performance and clutch and band durability. From a friction standpoint, an automatic transmission clutch consists of two basic elements: the friction-lined clutch plates and the steel reaction plates. As shown in Figure 2.1, the clutch plate assembly consists of three major components: a steel core, an adhesive coating, and the friction facings. The steel core is a part onto which the friction facings are bonded. The core is blanked from medium carbon steel with a Rockwell C hardness of 24 minimum. The core hardness ensures maximum tooth contact area. The minimum compressive load requirement can reduce the potential for wear or fretting on both spline surfaces. The friction facing is bonded to one or both sides of the steel core with an adhesive. The adhesive, a thermosetting organic resin, is selected to withstand high shear forces and a wide range of operating temperatures. The friction material must have the required friction characteristics for effective engagement, a pleasing (to the customer) shifting of gears, and durability. It also must withstand a wide range of operating temperatures as well as high shear forces and compressive loads. Friction material is produced from a variety of fibers, particle fillers, and friction modifiers. These materials, with properly blended composition, can provide the necessary material strength and bulk uniformity for successful manufacturing, as well as providing a heat-resistant material with the required friction properties. The porosity of the friction material allows the transmission fluid to be stored near the friction interface as a lubricant reservoir, and the resilience of the friction material permits it to conform to transmission mating surfaces. In automatic transmissions, multiple plate clutches, band clutches, and the torque converter clutches are used either to transmit torque or to restrain a reaction member from rotating. The function of an automatic transmission clutch and band is to serve as a lubricated brake (e.g., for a planetary gear element) to transmit torque between transmission elements and the torque converter pump (input) and turbine (output). Dynamic and static band and clutch friction torque levels, and the difference between static and dynamic friction torque, are important clutch performance criteria. They are closely dependent on transmission fluid characteristics. The transmission fluid can have a deep impact on shift-feel, chatter, static holding capacity, and the friction interaction between the transmission fluid and clutch material (Watts et al., 1990; Ward et al., 1993). All automatic transmission fluids recommended by vehicle manufacturers are formulated using a base stock plus a variety of additives, including friction-modifying additives to produce the desired transmission operation. Friction characteristics of the friction modifiers are described in Kemp et al. (1990) and Tung et al. (1989). Clutch friction characteristics for a base stock containing no friction-modifying additives are undesirable, as shown by the solid line in Figure 2.2. Although a high static coefficient of friction is desirable for good clutch-holding capacity, smooth engagement is difficult to achieve when static friction is greater than dynamic friction. As the clutch engagement approaches lock-up (decreasing sliding speed) with such friction characteristics, the increasing coefficient of friction tends to produce unstable stick-slip action between the rubbing surfaces and create abrupt and harsh clutch lock-ups. If these undesirable friction characteristics are too severe, shift-feel can be unsatisfactory to the vehicle's passengers.

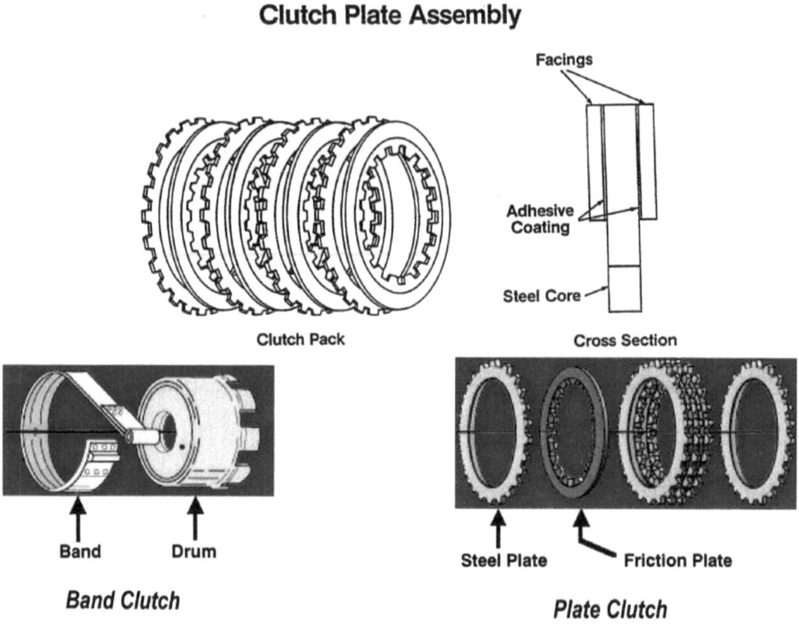

Clutch Plate Assembly

Figure 2.1 Band and Plate Clutches used to Shift Gears in Automatic Transmission

Objectionable shift-feel is a major cause of customer complaints about vehicle/transmission performance. The most desirable friction characteristics (Tung et al., 1989) for smooth clutch engagement are those described by curve A in Figure 2.2 (Kemp et al., 1990; Tung et al., 1989). High dynamic friction promotes a rapid clutch engagement, and static friction that is lower than the dynamic provides a smooth transition to clutch lock-up. However, the static friction coefficient must be high enough to provide good static clutch torque holding capacity. Friction-modified additives that produce properties described by curve B in Figure

2.2 are not entirely satisfactory. Although similar to curve A in shape, the low dynamic friction of curve B may demand increasing either the clutch area or the clutch applied pressure to complete a shift in a reasonable time. Under both dynamic and static conditions, clutch torque capacity is reduced. For curve C, friction characteristics are intermediate to those for curves A and B. However, due to unstable friction behavior for curve C, undesirable clutch lock-up and stick-slip action might occur for this case (Stebar et al., 1990).

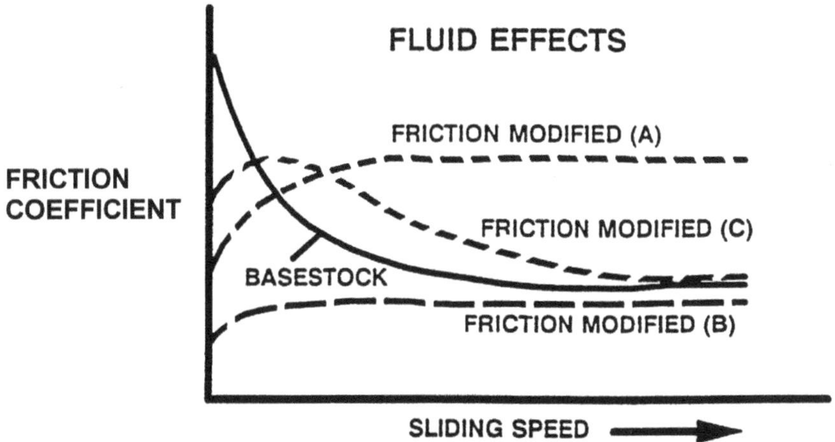

Figure 2.2 Automatic Transmission Clutch Friction Characteristics

The amount of clutch energy dissipated during clutch engagement can affect both transmission fluid and clutch friction material life. In a transmission power shifting clutch or band, torque capacity must be great enough to produce a rapid clutch or band engagement. Otherwise, the energy dissipated as heat can produce clutch surface temperatures that contribute to the deterioration of both clutch materials and transmission fluids. Typical clutch surface temperatures have been measured and found to be on the order of 95°C above sump fluid temperatures for wide-open-throttle conditions (Stebar et al., 1990). Figure 2.3 illustrates clutch energy dissipation for different transmission operating conditions; the clutch horsepower (clutch friction torque × clutch sliding speed) vs. engagement time is shown for three simulated heavy-throttle shifts. Curve A can be considered a satisfactory shift for which the clutch surface area is designed to provide long clutch life. Curve B illustrates an unsatisfactory clutch engagement because of extended engagement time. This curve indicates excessive clutch slippage resulting from an inability of the clutch to transmit sufficient torque to lock up the clutch in a reasonable period of time. If the energy can produce dynamic torques higher than the clutch can transmit, the clutch will not lock up, and the resulting energy will be so high that cellulose clutch materials are likely to burn or char. Such deterioration can occur in only a few severe clutch engagements. A surface temperature as high as 230°C above sump temperature has been reported for automatic transmission power shifts (Stebar et al., 1990). Curve C illustrates a clutch engagement for which the energy level is much different than for curve A, but the capacity of the clutch is high enough to produce a short engagement. Although of short duration, high clutch surface temperatures will deteriorate clutch durability due to severe energy dissipation. In summary, clutch and band friction characteristics, transmission fluid viscosity, fluid oxidation, fluid shear stability, and wear properties of friction materials all play a vital role in the overall performance and durability of the transmission. Because of the important characteristics of automatic transmission fluids in controlling transmission shift performance and long-term durability.

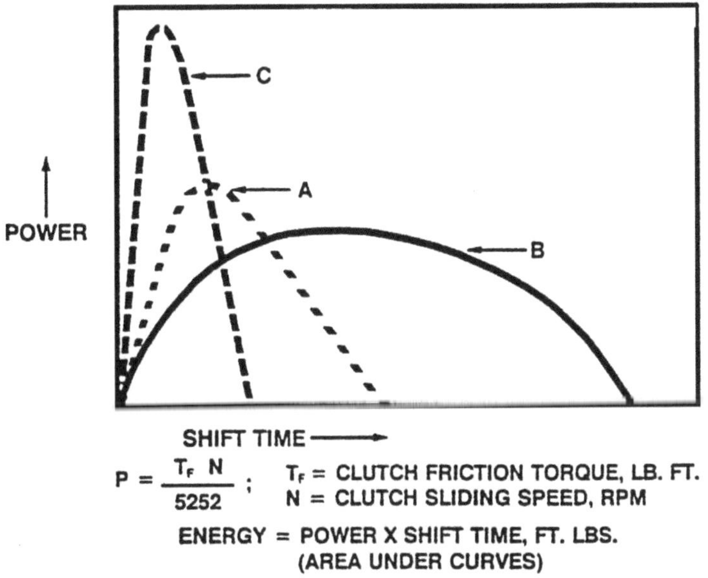

Figure 2.3 Automatic Transmission Clutch Energy Dissipation

2.2 Traction Drive

The development of the continuously variable transmission (CVT) by vehicle manufacturers has increased dramatically to take advantage of the fuel economy and better drivability benefits that can be achieved with traction drive components. A unique advantage of the CVT is that it allows an engine to operate over a range of speeds and loads independent of the torque requirements that are placed on the wheels by the vehicle and the driver (Heilich, 1983). Traction drive CVT was applied to vehicles in the 1930s and was started by the Toric transmission (Hewko et al., 1962). Three basic types of commercial CVTs (Kluger et al., 1997) have been developed, as follows: (1) belt (steel, fabric, push, and pulley type); (2) half or full toroidal traction; (3) epicyclic. As indicated in Figure 2.4, two commercial CVTs (belt-pulley and traction drive) are the most common CVT drive units. Production vehicles equipped with steel belt pulley CVTs have been in the automotive market for more than 10 years. However, the full potential for reduction of fuel consumption could not be realized until the fully integrated electronic control system became available. The belt-pulley CVT relies on a metal link belt developed by Van Doorne's transmission and is typically limited to use with engines of 2.0-liter displacement or smaller (Hendriks, 1993). Around 1990, a full toroidal traction drive device was developed to supplement the belt-pulley type because of the traction drive's ability to handle larger engines. The toroidal traction drive provides better mechanical efficiency because of a very efficient rolling motion that operates in the Elastohydrodynamic regime (Dowson et al., 1991). The current traction drive transmission transmits power through a rolling contact via forces that are varied with the radius at which the traction force is applied. Thus, power is transmitted in a continuous manner. This concept relies on two rolling elements that place a thin film of lubricant into an extreme shearing condition. The thin film between the two rolling bodies experiences high load and extreme shearing force, which produces complex Elastohydrodynamic phenomena. The viscosity of the traction fluid increases significantly as a consequence of the high pressure, and the fluid becomes almost glasslike. The tangential force that is transmitted between the driving and driven elements is greatly enhanced by the traction fluid. The surface contact area between the two rolling elements is a finite area whose orthographic projection is an ellipse (Kluger et al., 1997). Within the ellipse, fluid is trapped between the rollers and is subjected to high compressive stresses of 0.7 to 3.5 G Pa (100,000 to 500,000 psi). These stresses increase the instantaneous viscosity of the fluid by several orders of magnitude. This semi-solid lubricant can then transmit torque through the drive. Basic research was conducted with application to different traction drive units (Heilich, 1983). However, a traction drive CVT was not applied in practical use until 1990, due to three major barriers (Machida et al., 1990):

1. The traction drive rolling element materials were not sufficiently reliable due to high temperature and high pressure at the traction contact point.

2. There were no bearings that could support high speed and large axial load.

3. There were no traction fluids that could meet all traction component requirements.

A traction drive requires not only higher traction coefficient to meet power transmission requirements, but also better lubrication and hydraulic lubricity requirements. Traction elements are generally made of materials similar to those used for ball or roller bearings for which the surface must withstand high loading and long duty cycles. This requirement also demands hardened steel or hardened materials for the contact surfaces. Because the traction drive operates at high speeds, it therefore provides high input and output ratios. Advanced tribological materials in the rolling element have made the life of the traction contact point long enough for use in high-power automotive transmissions. The efficiency of the traction drive relies heavily on the lubricant that is used. A high traction coefficient is required even at high contact temperatures. The temperature of a typical traction drive can rise to over 140°C. In addition to the traction requirements at high temperatures, traction fluids must not deteriorate when subjected to repeated shear or oxidation conditions. It has been indicated by traction fluid researchers (Tsubouchi et al., 1990; Hata et al., 1998) that naphthenic base compounds and several types of synthetic fluids have better traction characteristics than paraffinic or aromatic compounds. Research on traction fluids has also indicated that the traction coefficient of naphthenic compounds provides a higher rotational barrier and has a stronger molecular stiffness compared to paraffinic or aromatic compounds (Machida et al., 1990). Other variables affecting traction properties include rolling speeds, lubricant temperature range, quality of the roller surface finish, Hertzian contact pressure, degree of spinning, and the geometry of the rolling bodies. A substantial amount of CVT research is focused on traction fluid formulation and evaluation to determine whether the fluid can meet traction drive system performance requirements. Future research will concentrate on traction fluid formulation to optimize overall system performance and to adapt the traction fluids to the target automotive applications.

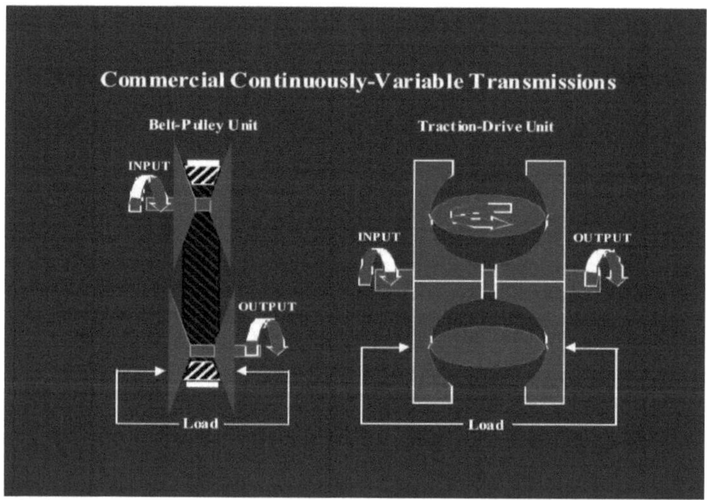

Figure 2.4 Belt-pulley and traction drive units for continuously variable transmission

2.3 Universal and Constant-Velocity Joints

Universal joints and constant-velocity joints are used to transmit power from the main driveshaft to the vehicle's wheels, as shown in Figure 2.5 (Nunney, 1998). The top portion of the figure is a modern bipot universal joint, and the bottom figure portion is an application of constant-velocity universal joint for front-wheel drive. There are several different designs for joints. All have the ability to transmit power between two shafts with some degree of axial misalignment. The joints may be of the fixed or plunging type. The plunging type incorporates a spline coupling to allow axial approach of the shafts. Typically, the joints consist of an array of rolling elements held by a retaining cage between two raceways. The entire assembly is packed with grease and sealed inside a rubber boot. The contacts within the joint are between the rolling elements (balls, needles, or rollers), the raceway grooves, and the cage. The rolling elements are usually made from a high-carbon steel that is through hardened. The raceways are made from a medium-carbon steel, and the load bearing areas are induction hardened or carburized. When the shafts are aligned, the rolling elements are stationary. When the shafts are misaligned, the rolling elements undergo a low oscillation reciprocating motion. The load on the ball race contact is high, and the entrainment velocity is low. Because the surfaces tend to be relatively rough, boundary lubrication is the dominant lubrication regime. Possibly the most common form of failure in the field is by damage to the boot. This results in loss of lubricating grease and failure by gross damage to the joint components. However, contact fatigue is also observed. Cracks initiate at the surface and propagate to form spalls in the near-surface region. Extended use of the joint results in wear of the balls and raceways, leading to loss of dimensional accuracy.

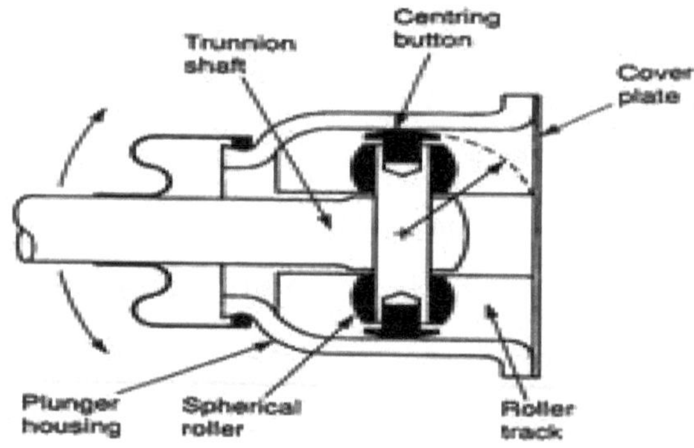

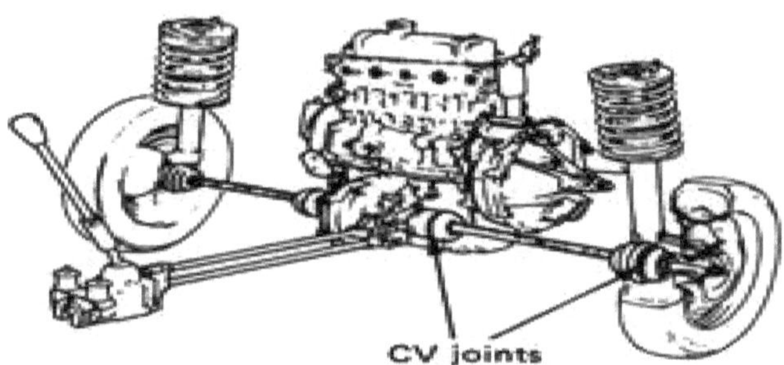

Figure 2.5 A Modern Bipot Universal Joint (top) and Application of Constant-Velocity Universal Joint (bottom) for Front-Wheel Drive. (From Nunney, M.J. (1998), Automotive Technology, SAE International, Butterworth-Heinemann)

2.4 Wheel Bearings

Conventional rolling element bearings are used in the road wheel hubs. These standard manufactured components are hardened high-carbon steel, machined and ground to a high tolerance and surface finish. Usually, deep groove bearings or a back-to-back angular contact bearing assembly is used. The bearings are packed with grease on assembly and sealed with elastomeric lip seals. The most common cause of wheel bearing failure is by damage to the surface caused during incorrect fitting or following impact during operation (the wheel striking a curb). The surface damage acts as an initiation site for rolling contact fatigue failure.

2.5 Drive Chains

Some automotive engine designs use chain drives for timing and driving ancillary components. Chains have the advantage of a high capacity, increased durability, and being relatively cheap and simple. The basic components of a chain drive are the chain, sprockets, chain guides, and tensioners. There are two main types of drive chain: (1) roller chains consist of link plates with a pin, bush, and roller assembly; and (2) silent chains consist of an array of shaped link plates and pivot pins. The chain is lubricated by two means, firstly by splash as it dips into the sump during its travel and secondly through jets that impinge onto the sprocket/chain entry. The rollers in the case of roller chain, and the link plates in the case of silent chain, locate in the teeth of the sprockets. In the roller chain, the pin slides inside the bush during articulation, while the roller rolls across the sprocket tooth. The pin bush contact can therefore undergo excessive wear, causing chain elongation. For this reason, the pins and bushes are usually case-hardened and ground smooth, while the link plates are usually unhardened steel. With silent chain, the link plate slides across the sprocket tooth; this can result in wear of the link plate surface. The chain drive is subjected to dynamic loading. Chain guides and tensioners are incorporated to reduce the associated vibration. The guides are generally made of a polymer (typically Nylon 66) and either bolted in place or loaded against the chain by a spring or hydraulic tensioner. The chain tension and impact forces between the chain and guides can result in wear of the guide faces. Wear debris or dust in engine oil can become embedded in a polymeric tensioner and can act as an abrasive agent on the timing chain. Fuel components or fuel reaction products can enter engine oil and may soften a polymeric tensioner. Wear occurs on the guide face where it contacts the link plate edge, forming "tramlines." In some cases, this continues until the grooves formed are deep enough that the rollers make contact with the guide; the load is then distributed to a greater extent and wear is reduced.

Chapter Three
The Tires

3.1 Introduction

The word "tire" is derived from "attire" a protective cover to contain compressed air (or sometimes gas) in a toroidal chamber that is attached to a steel rim or wheel of an automobile. The North American spelling (tire) is closer to attire; the British spelling is tyre. Typically the tire has to perform the following functions.

3.1.1 Transmission of Forces from the Vehicle to the Road during Drive:

These forces include the weight of the automobile, passengers, and the cargo; the forces required to accelerate and decelerate the automobile; and the forces needed to steer the vehicle. These forces need to be transmitted in such a way that wear and fatigue failure of both the tire and the pavement are within acceptable levels. This is done effectively by providing a large contact area between the tire and the pavement, leading to a low contact pressure.

3.1.2 Provision of a Firm Grip on the Road:

The tire must provide a firm grip on the road that may be dry or wet, and sometimes even contaminated intentionally (salt during winter) or accidentally (spillage). This is done by choosing an appropriately vulcanized rubber, to provide a large coefficient of sliding friction at the tire/pavement interface. The associated but rather conflicting demand on the tire is that the values of rolling friction and hysteresis loss should be as low as possible. Pneumatic tires are critical components of an automotive vehicle about 14% of all fatal accidents occur when roads are wet

(Pottinger et al., 1986) and the grip is lost, which is typically only about 3% of the driving time.

3.1.3 Absorption of Unevenness of Roads and Provision of Comfortable Journeys:

The tire is also used to absorb unevenness of the road and provide a comfortable journey. The texture or the roughness characteristics of the pavement and the tire become important in determining the quality of ride and the noise that is produced by micro-slip in the contact area.

Obviously, the tire has to do this at a low cost and must have a sufficiently long life with ample safety provisions against accidental deflation. It is not uncommon for truck tires to last about 160,000 kilometers (about 100,000 miles) and car tires about 80,000 kilometers (about 50,000 miles) (French, 1988). These lives can be increased by retreading.

3.2 Construction of a Tire

In the early days, tires generally with toroidal tubes that were inflated with pressurized air, were cemented to the wheels and sometimes even fixed with rudimentary holding devices. This all changed with a patented design that incorporates rather inextensible steel wire hoops (tire beads) in the inside of the tire casing to fix and locate it on the wheel rim accurately, Figure 3.1a. The central depression in the rim facilitates tire fitting and its removal (Figure 3.1b). Subsequent tires were of cross-ply construction (Figure 3.2a), similar to the original produced by Dunlop in 1888. Michelin, in 1948, started producing tires with radial plies (Figure 3.2b). Both types carry patterned rubber treads and are used in different circumstances. The behavior of a particular tire in terms of load-carrying capacity, life, steering characteristics, resistance to damage, ride quality, and noise is dependent on the tire casing and is affected by the rubber/plies layout. In general, cross-ply is used in military applications, aeronautical, agricultural, cycle, and motorcycle tires; and radial ply is used for most passenger cars and trucks (French, 1988). Some of the common patterns for the rubber tread are shown in Figure 3.3. The purpose of these patterns is to squeeze water out of the contact in wet conditions and avoid skidding or hydroplaning. Feeder channels or smaller grooves flush water out of the contact. The grooves are generally 3 mm wide and 10 mm deep for a new tire (Moore, 1975). The tread between these grooves may be provided with cuts or sipes (small hook-shaped or bracket-shaped grooves in the tread of an automobile tire) to facilitate a wiping action to remove thin water films. Tread patterns, especially on the sides, help in cooling the tire by circulating the air.

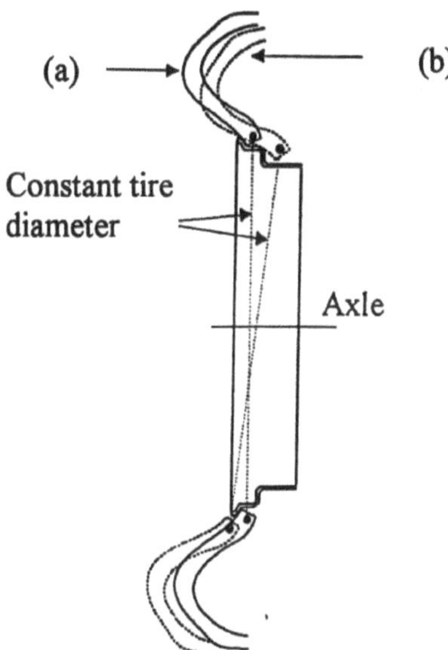

Figure 3.1 (a) Normal Operating Positions of Tire. (b) Tire Position during Fitting and Removal

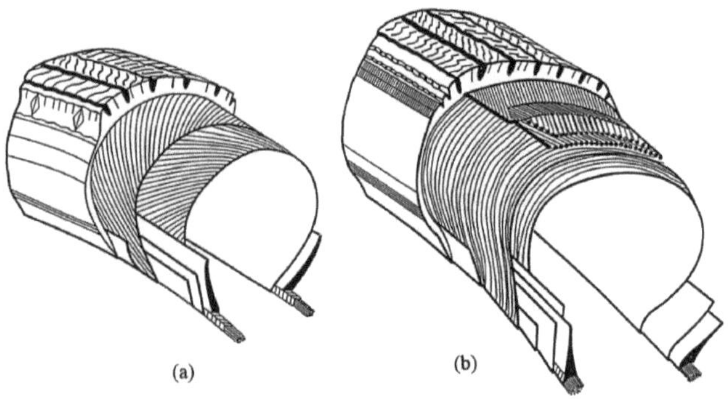

Figure 3.2 Construction of a Tire: (a) Cross Ply; (b) Radial Ply. (From Clark, S.E. (1971), The Contact between Tire and Roadway, in Mechanics of Pneumatic Tires, Clark, S.E. (Ed.), National Bureau of Standards Monograph 122)

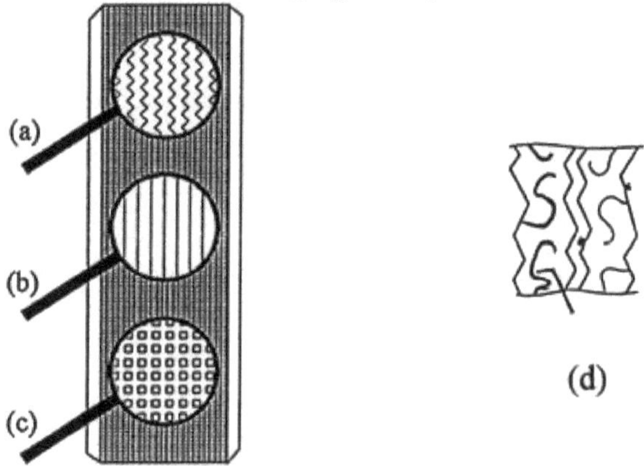

Figure 3.3 Tire Tread Pattern: (a) Zig-Zag; (b) Ribbed; (c) Block; (d) Sipes in Tread

3.3 Mechanics of Load Transfer

Consider a cycle wheel: the hub is connected to the rim by wire spokes. An upward force at the bottom of the rim is transferred to the hub by the tension of the spokes. Load transfer from an automobile wheel to the tire is analogous to that of the cycle wheel. The wheel behaves like the cycle hub, and the plies act like spokes. Pressurization of the tire produces tensile hoop stresses around the circumference and also along the direction of the plies. The tension in the plies is transmitted to the tire bead and then to the wheel rim. Application of a vertical force develops a contact area with the pavement, and the tire deforms near the contact. The angle at which plies are pulling the tire bead increases, and the component of tensile force in the radial direction decreases. The radial pull in other directions remains approximately the same, and the net result is a vertical force on the tire bead that gets transmitted to the wheel (French, 1988). Note that if the tire is rigid, a system of stresses will develop to transfer the load, as in the case of railroad wheels. Figure 3.4 shows a wheel rolling on the pavement and carrying a normal load W. The rolling resistance is shown by the resultant force F_r. In the case of free rolling, the wheel has no applied moment, and therefore the reaction of the contact forces passes through the wheel center. In the case of a braking or a driving wheel, the presence of a turning moment implies that the resultant force does not pass through the center.

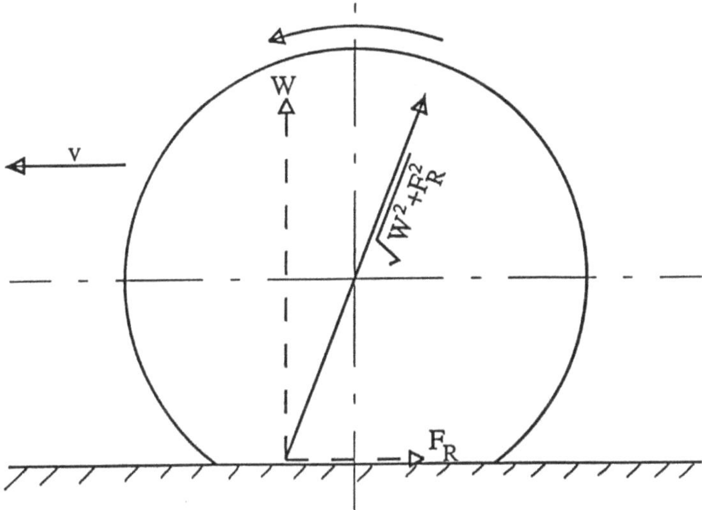

Figure 3.4 Load Transfer in a Tire

3.4 Contact Area and Normal Pressure Distribution

The tire tread, which is fairly inextensible and loaded by the hoop tensile stresses, contacts the pavement and changes its curvature to drape over the surface. A fairly large contact patch is generated, the size and shape of which depend on tire pressure, tire construction, and applied load. Some typical contact patches are shown in Figure 3.5. A typical normal contact pressure distribution for a tire with no tread pattern is shown in Figure 3.6. A typical value for average contact pressure is about 1 MPa. By comparison, steel wheels and steel rails in railroad applications produce an average contact pressure of about 1000 MPa. The presence of a tread pattern reduces the real contact area and increases the peak contact pressure.

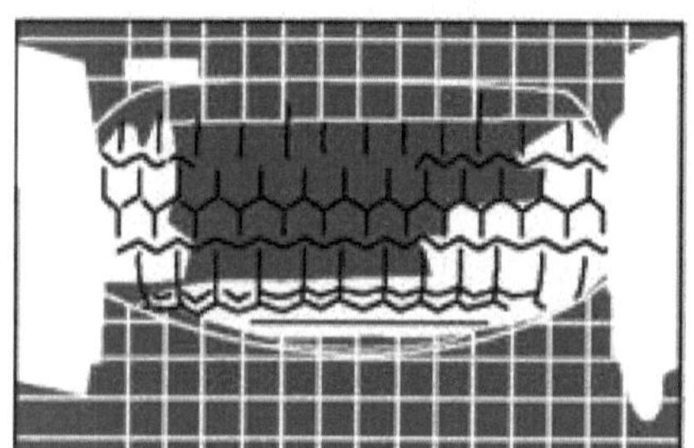
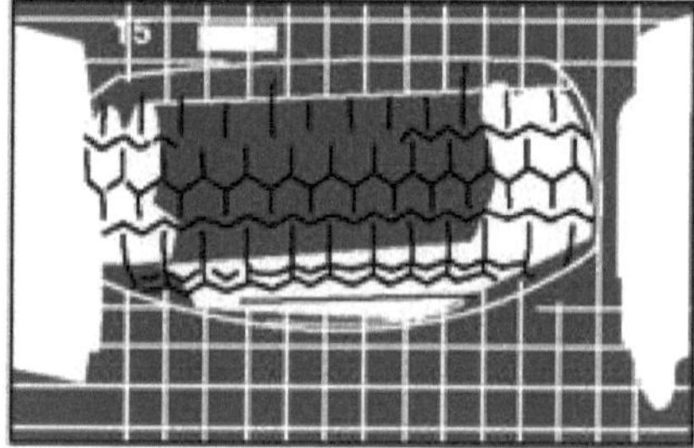
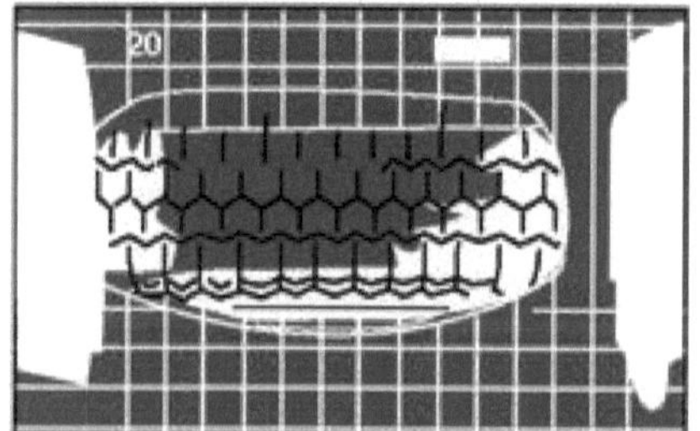
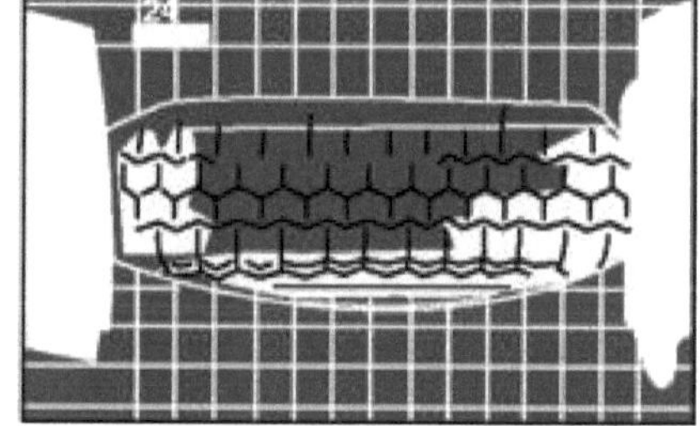

Figure 3.5 Tire Contact Area at Increasing Tire Pressure (Clockwise from Top left corner). (From French, T. (1988), Tyre Technology, Adam Hilger)

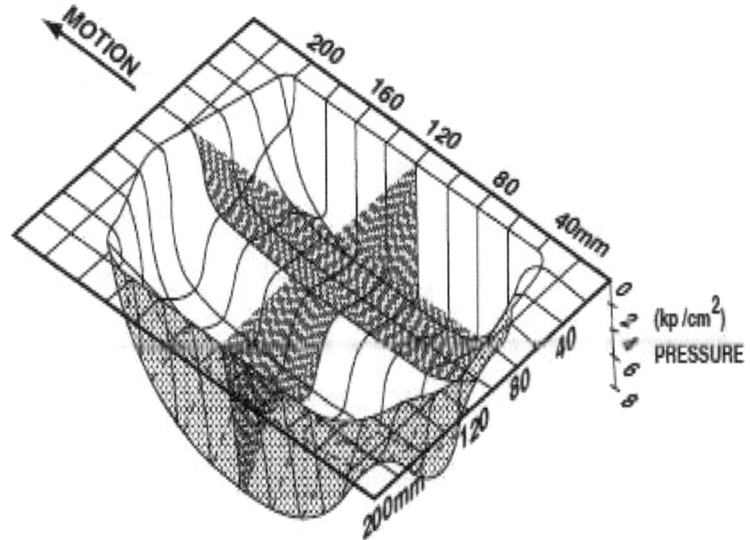

Figure 3.6 A Typical Contact Pressure Distribution. (From Clark, S.E. (1971), The Contact between Tire and Roadway, in Mechanics of Pneumatic Tires, Clark, S.E. (Ed.), National Bureau of Standards Monograph 122)

3.5 Slip and Generation of Shear Traction

The difference between the elasticity of the tire and the pavement material results in a difference in tangential displacement within the contact area, leading to a tendency to slip. Friction forces that are generated locally resist the slip. In some areas, known as stick regions, the local friction is sufficiently large to resist slip; but in other areas, known as slip regions, the friction force reaches its limiting value (equal to the load multiplied by the coefficient of sliding friction at that point) and local slipping occurs. Figure 3.7 shows stick and slip regions for a tire/pavement contact under normal pressure and stationary conditions. However, if the tire were to roll, whether braking or driving, the stick region shifts to the front of the contact (Figure 3.7b). In a braking situation, the slip direction is toward the stick region; whereas in a driving situation, the slip direction is away from the stick region. During cornering, the behavior is complicated by sideways slip (Moore, 1975; Clark, 1971); a typical contact patch is shown in Figure 3.7c. In the case of braking wheels, just before the tread enters the contact zone, it is under a tensile stress, and it elongates. In the stick zone, the tread and the pavement have no relative motion, and it is only in the slip zone that the tread slips toward the stick zone and starts to contract. In the case of driving wheels, the tread is under a compressive stress and goes into the contact compressed. It recovers its compression in the slip region by slipping away from the stick region. The distribution of shear stresses in the contact area for free rolling, braking, and driving are shown in Figure 3.8. The shear stress distribution, as well as the normal pressure distribution, depends on the construction of the tire, inflation pressure, and several other factors. Because of the decreased (or increased) tread length in the stick region of the contact, the apparent wheel circumference is too short (or too long) for estimating the distance traveled. In one wheel rotation, the braked wheel travels a larger distance than its circumference, whereas the driving wheel covers a smaller distance. The brake slip ratio S_B and the drive slip ratio S_D are defined as (Moore, 1975):

$$S_B = \left(\omega_{Ro} - \omega_{br} \right) \Big/ \omega_{Ro}$$

$$S_D = \left(\omega_{dr} - \omega_{Ro} \right) \Big/ \omega_{dr}$$

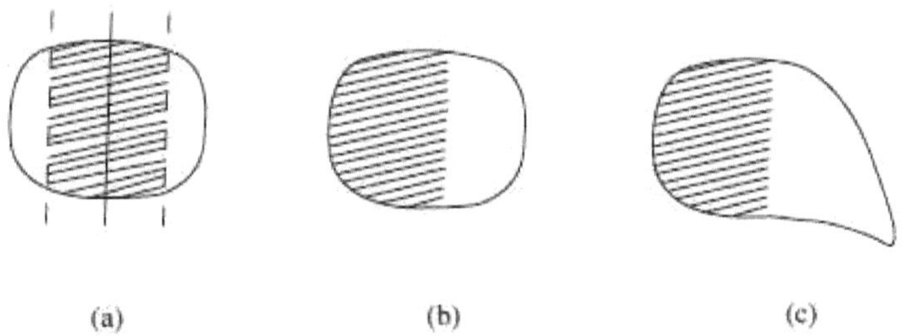

Figure 3.7 Stick and Slip (Shaded) Regions when the Vehicle is (a) Stationary; (b) Braking or accelerating; and (c) Cornering

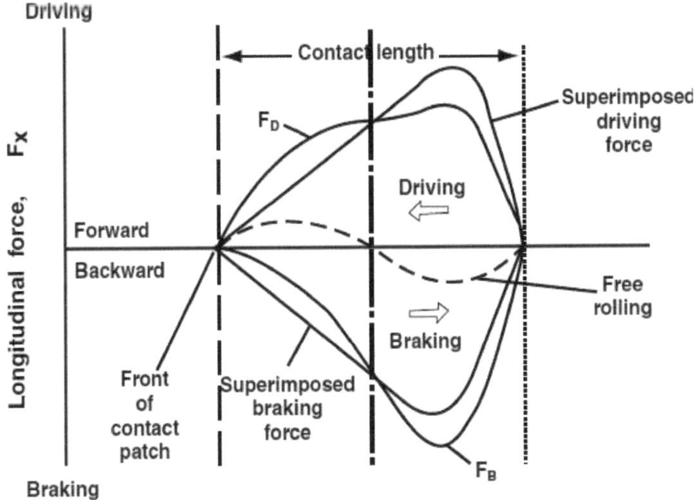

Figure 3.8 Distribution of Shear Traction within the Contact Area. (From Moore, D.F. (1975), Principles and Applications of Tribology, Pergamon Press)

Where ω_{Ro} is the angular velocity of a wheel in free rolling, and ω_{br} and ω_{dr} are the angular velocities during braking and driving, respectively, for the same forward speed. Typical variations of braking force and driving force normalized with respect to the normal load are shown in Figure 3.9. Finally, being toroidal, the tire surface is non-developable; that is, it cannot be manufactured from an un-stretchable plane surface like a cylindrical surface can be (Clark, 1971). Contact with a flat pavement therefore requires the tread to undergo uneven local stretching, a phenomenon known as squirming. This movement is, however, very small in comparison to the slipping described above.

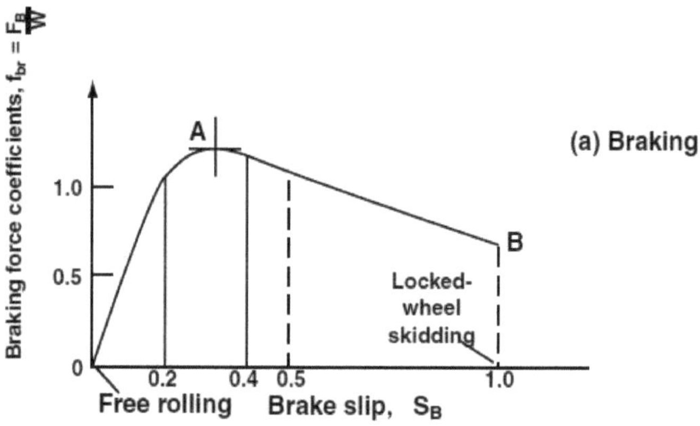

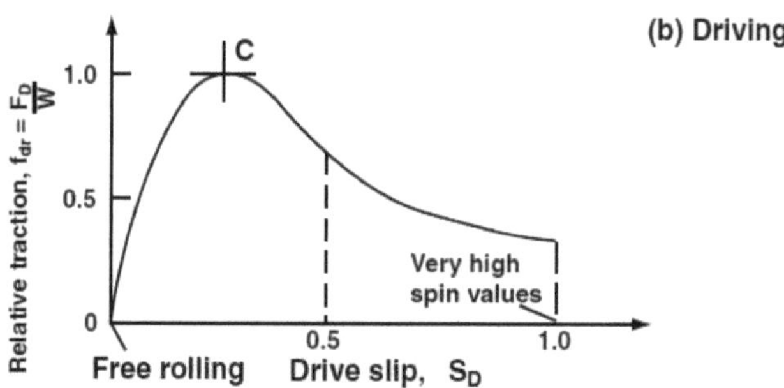

Figure 3.9 Variation of (a) Braking Force, (b) Driving Force, with Slip. (From Moore, D.F. (1975), Principles and Applications of Tribology, Pergamon Press)

3.6 Hydroplaning

In wet conditions, a fluid film can form between the tire and the pavement, as shown in Figure 3.10. In Zone 1, the fluid film is complete and there is no contact between the tread and the pavement. The tire experiences virtually no friction force in this zone. In Zone 2, the fluid film thickness gradually reduces to nearly zero. Here, there is partial contact between the tire and the pavement, but the contribution to friction is very small. In Zone 3, the fluid is completely squeezed out, and intimate contact occurs. It is only this zone that provides sufficient frictional traction for driving, braking, and steering. Under certain conditions of high speeds or thick water layers on the pavement, Zone 3 may be absent. The tire then rides on a film of water, and the condition is known as hydroplaning. Because the traction offered by a water layer is negligible and can provide only small braking or steering forces, this condition is dangerous, and the tire is provided with tread patterns so that water is squeezed out to the front and to the side of the contact.

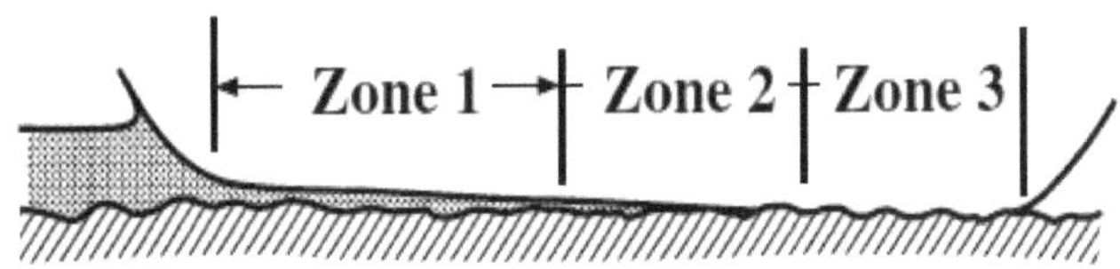

Figure 3.10 Hydroplaning: The Contact Area is divided into Three Zones

3.7 Wear Characteristics

When in service, the tire tread is subjected to many complex events. It deforms and undeforms as it goes through the contact and drapes over pavement features. This produces a fatigue-type loading leading to failure. As shown in Figure 3.7, the tread slips over the pavement and endures abrasion by harder particles either buried in the pavement or present in the contact as third bodies. This manifests itself in the form of tears, cuts, and scratches. Sometimes, wear by roll formation may also be present, and the situation is complicated by chemical degradation and temperature effects (e.g., see Veith, 1986). Abrasion by slipping in the contact patch contributes most to the tire wear, and any maneuver that increases slipping increases wear. Cornering, swerving, and even excessive braking and accelerating increase the wear rate considerably. Some of the common terms used to define tire wear are:

3.7.1 Tread loss:

The averaged value of the depth loss of tread through wear. It is measured for all the grooves and expressed in millimeters. Most countries have legal limits on the minimum tread depth for safety reasons.

3.7.2 Rate of wear:

The loss of tread in millimeters per 10,000 km of travel. Typical values are a few millimeters per 10,000 km (Moore, 1975).

3.7.3 Tread wear index:

The ratio of wear rate of a control or reference tire to that of the experimental tire, expressed in percent.

As is usually true for wear of any component, the tire wear is also influenced by the two contacting bodies (i.e., the tire and the pavement), the interface, and the operating conditions. Various influencing parameters include the tire design and the pavement design, their construction, the composition of the tread, load on the tire, inflation pressure, speed, the manner in which the vehicle is being driven, and ambient temperature and other weather conditions. Because the variability in these is generally larger than that in other wear situations, a correlation between laboratory tests and field tests is very difficult. Even in field tests, differences exist in values obtained using fleet tests (driven cars) vs. trailer tests (towed cars) (Pillai, 1992).

Chapter Four
The Brake System

4.1 Introduction

Slowing down an automobile and/or permitting movement with a constant speed on gradients require dissipating kinetic and potential energy of the vehicle by a braking action. It is possible to convert this energy into rotational kinetic energy (spinning a flywheel) and reuse it later in accelerating the automobile, but the considerable weight and cost associated with regenerative braking systems limit their use to only a very few energy-conscious vehicles. Another method for slowing a vehicle involves dissipating this kinetic and potential energy by means of friction, which is the most common method of braking in passenger cars, light and heavy trucks, buses, and off-road vehicles. The generic systems used are shown in Figure 4.1. In a drum-type brake (Figure 4.1a), the shoe applies normal force on approximately 50 to 70% of the drum circumferential area. The drum and shoe are made of high-friction, low-wear materials, and the frictional loss at the interface provides dissipation of energy and the necessary braking action. In general, the brake shoe is the sacrificial component and wears faster than the drum, thus requiring comparatively frequent replacement. In a disc-type brake (Figure 4.1b), approximately 7 to 25% of the disc-rubbing surface is loaded by the brake shoe. Disc brakes provide relatively better heat dissipation because they have larger exposed surface areas and a better cooling geometry. However, the exposed surface area makes them vulnerable to unwanted contamination. Due to cooling characteristics, contamination, and other design issues, front brakes are of the disc type and rear brakes of the drum type

50

(Anderson, 1992). Apart from the interface between the brake-lining and the brake-drum/disc, energy is also dissipated at the tire pavement interface, and it is the optimum combination of the two interfaces that produces effective braking. For example, if the road is icy, braking action will lock the wheel, and the tire will slip or skid on the road. The friction force will be very small, and braking will not be effective. A similar situation will arise if the brake-lining–brake-drum/disc interface becomes contaminated, and the friction generated at this interface becomes very small.

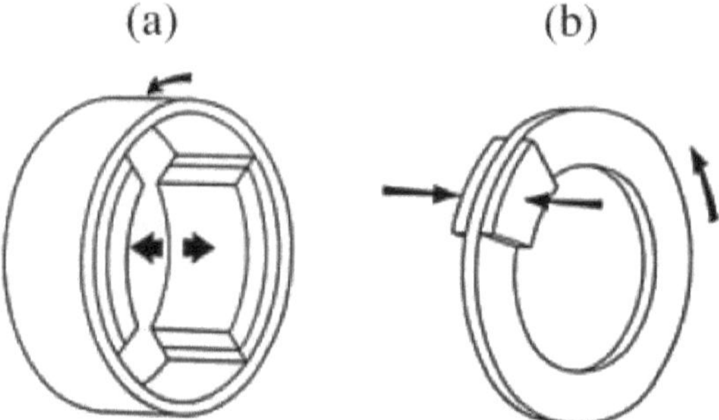

Figure 4.1 (a) Drum Brake; (b) Disc Brake

4.2 Contact Pressure and Shear Traction

Figure 4.2a shows a brake shoe and disc in contact. The applied force generates a distribution of normal contact pressure at the interface. When the brake is new, the applied normal pressure is constant over the entire contacting area, as shown in Figure 4.2b. Sliding generates a distribution of frictional shear traction; the value of local shear stress equals the coefficient of sliding friction multiplied by the local normal pressure. The rate of energy dissipation is equal to the sliding velocity multiplied by the frictional shear stress. Because the sliding velocity increases linearly with the radial distance of the point from the center of the brake disc, more energy is dissipated at outer radii than inner. Wear at various points can be considered to be proportional to the rate of energy dissipation, and so it is larger away from the center. Under steady-state conditions, however, wear occurs uniformly over all the rubbing surface and requires that the energy dissipated at every point be constant. This means that the contact pressure multiplied by sliding velocity is constant, or that the contact pressure follows a 1/ (radial distance) relationship (Figure 4.2c). As expected, the contact pressure at the ends gently drops to zero. Note that this mechanism is self-equilibrating; and, thus, if the local conditions change due to material inhomogeneity, contamination, or for any other reason, the contact pressure will change appropriately to produce a uniform wear rate. The reasoning given above assumes that the surfaces are smooth and ignores the fact that the surfaces are rough and contacts are made at asperity summits. The contact pressure is not smooth as depicted in Figure 4.2, but will have isolated peaks. However, the argument of uniform wear rate still applies.

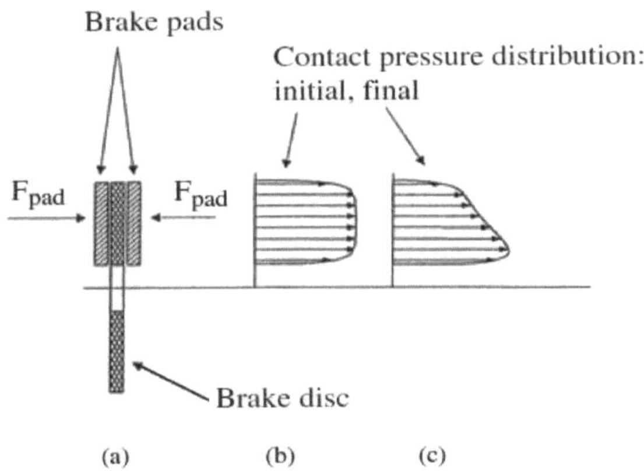

Figure 4.2 Variation of Contact Pressure in a Disc Brake

4.3 Materials

Brake material must meet conflicting requirements: it must combine a high friction coefficient with a low wear rate. The large amount of kinetic and potential energy dissipated as frictional loss heats up the brake lining to a few hundred degrees Celsius (Moore, 1975). The asperities can reach a flash temperature of 1000 to 1100°C (Anderson, 1992). The demands on the brake materials are many and great, and the industry relies on highly proprietary formulations. Broadly, these can be classified as organic, metallic, and carbon (Anderson, 1992). Carbon-based brakes such as carbon-carbon composites are generally used for aircraft and racing car applications wherein weight is a critical design issue. Metallic brake linings, mostly of copper, iron, sintered bronze, and mullite, are used for very high power input density applications such as high-speed railways and racing cars. Cast iron is also used in some applications, but nowadays predominantly used as a counter face material in automotive drum and disc brakes.

4.4 Friction

Figure 4.1 shows the most commonly used brake systems. Application of force on the brake shoe produces the required frictional torque. The frictional torque depends on the coefficient of friction between the brake shoe and the drum/disc and on many design parameters. In the case of drum brakes, the mechanism of self-actuation results in a nonlinear relationship between the frictional torque and the coefficient of friction. To overcome this, the frictional performance of a brake is characterized by Brake Effectiveness, which is the ratio of the brake frictional torque to the applied force. Figure 4.3 shows Brake Effectiveness for various drum and disc brakes. Surface irregularities and the cumulative effect of manufacturing tolerances in a new brake system mean that the contact is not made over the entire rubbing surface area. Only after some wear and burnishing (plastic flow of asperities) can the contact take place over most of the intended area. This results in the early Brake Effectiveness (Green Effectiveness) being lower than the steady-state value (Burnished Effectiveness). Brake Fading refers to dropping of the Brake Effectiveness to low values and results from excessively high temperature, migration of resin due to cooling, blistering of the brake material, and contamination. Brake Effectiveness is sensitive to speed and generally drops with increasing speed. Environment also has an effect. Contamination in the form of water, oil, dust, and/or corrosion products can change the effectiveness, and braking systems are designed so as not to be adversely affected by these factors.

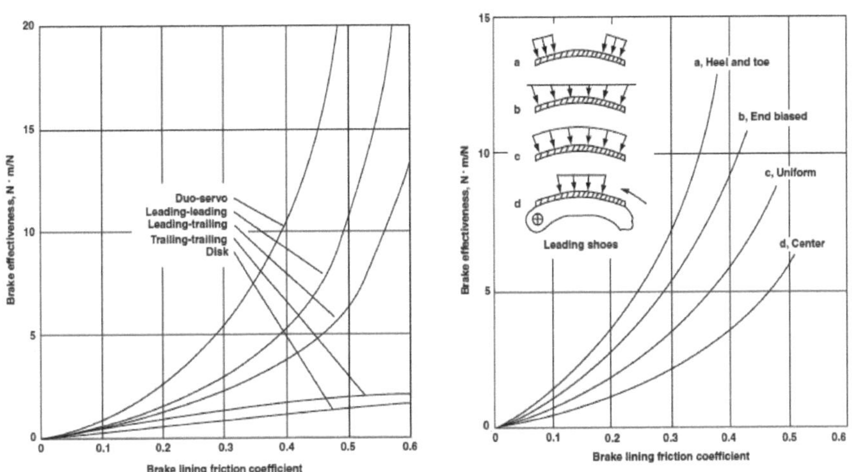

Figure 4.3 Brake Effectiveness vs. Brake Lining Friction Coefficient. (From Anderson, A.E. (1992), Friction and Wear of Automotive Brakes, in ASM Handbook, Vol. 18, Blau, P.J. (Ed.), 569-577)

4.5 Wear

Brakes encounter the following wear mechanisms: thermal, abrasive, adhesion-tearing, fatigue, and macro-shear (Moore, 1975). As mentioned, very high flash temperatures (up to 1100°C) are reached at asperity junctions, and the brake material may melt or thermally decompose, leading to loss of material. Oxidative wear also occurs near the edges. Abrasive wear occurs due to plowing by wear debris and hard contaminants. Adhesion and tearing occur due to brake material sticking to the counter face and tearing off. Fatigue wear occurs due to repeated asperity encounters; and macro-shear is the sudden failure of friction material that has been weakened by heat. For practical purposes, brake wear is generally defined in terms of the distance traveled or the usage time. It can also be measured in mass worn per unit of frictional work. Typical values range from 40 to 100 mg/MJ (Anderson, 1992). During burnishing or break-in or running-in, wear of both the brake material and the counter face (drum or disc) is high. After burnishing, the wear stabilizes at a lower rate.

Chapter Five
Windshield or Windscreen Wipers

Windshield (or windscreen) wipers have a slender strip of a rubber compound, called the blade, supported by 4, 6, or 8 clips, as shown in Figure 5.1. The arm applies a load of 1 to 2 kg (10 to 20 N) to the clip hinge center, and this is distributed fairly uniformly along the length of the blade (Figure 5.2). The primary function of the wiper is to improve visibility by removing the dirt particles on the windshield by a wiping action. This is usually accomplished with the windshield washer on. The chemicals in the fluid help in loosening dirt particles, which are then scraped by the blade. If it is raining, the wiper removes excess water and leaves a thin and even film of water on the windshield. If there is only a little water on the windshield, the wiper spreads it evenly and improves visibility. Typical sliding speeds are 50 to 100 cm/s, and the coefficient of friction ranges between 0.15 and 0.2 for the wet case (rain water or with windshield washer fluid) and about 1.4 for the dry case. The friction coefficient in the wet case exceeds the value that would be expected if a hydrodynamic film was formed. It is generally believed that the wiper operates in the mixed lubrication regime, that is, some asperities touch the glass windshield. These asperities are formed by carbon filler material, which is incorporated in the blade to improve its wear performance. This intermittent contact leaves a thin stream of water behind, which then breaks into little droplets, reducing visibility. Sometimes, uneven wear of the blade can result in certain areas of the windshield not being wiped with sufficient contact pressure, leaving behind an unclean surface. Various tribological tests for windshield wipers include running the wiper on a rotating flat glass disc, running it on a sample windshield, and also in field tests. During field tests or wind tunnel testing, air flow over the wiper generates a lift and reduces the contact pressure. This leads to a deterioration in wiping performance and produces a realistic test of the wiper performance.

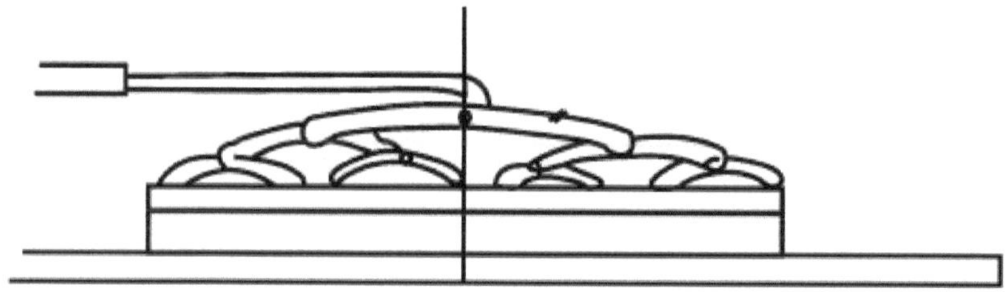

Figure 5.1 Construction of a Typical Wiper

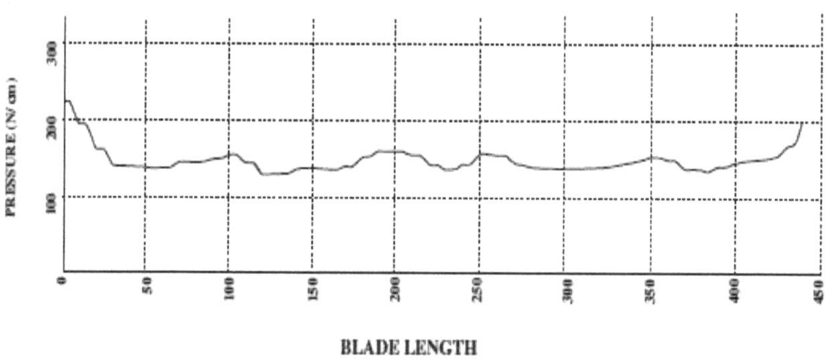

Figure 5.2 Contact Pressure Variation along the Blade Length

Chapter Six
Automotive Lubricants

This chapter deals only with those substances that act as lubricants in cars and trucks.

6.1 Lubricant Function

An automotive lubricant provides some or many of the following beneficial functions. It forms a fluid film that keeps one moving surface out of direct contact with the opposing surface. Additives in the lubricant are designed to lay down a chemical film that provides wear protection, particularly while operating under boundary lubrication conditions. Oil additives also help slow the rate of deposit, sludge, and varnish formation and slow the rate of corrosion, rust, and chemical attack on components. The circulation of the lubricant which transfers heat away from hot surfaces, warms cold surfaces, and transports debris and contaminants to the filter. Oil additives that reduce friction contribute to the energy efficiency of the vehicle. Some automotive lubricants increase traction, control stick slip, and assist in sealing by providing a fluid barrier to reduce the escape of gases and by supplying chemical agents that provide a small amount of seal swell. In addition to performing these valuable functions, the lubricant must remain stable for an extended period of time and in some cases must act as a working fluid for such engine components as hydraulic lifters or chain tensioners.

6.2 Lubricant Types

Engine oil and automatic transmission fluid are two of a vehicle's lubricating fluids that provide many of these functions. Other substances in a vehicle that perform at least one lubrication function include air conditioner fluid and its associated lubricant, gear and axle lubricants, grease, brake fluid, power steering fluid, and shock absorber fluid. In addition to the above fluids that are commonly recognized as lubricants, several additional fluids also provide some lubrication functions. For example, fuel helps lubricate the fuel pump, injectors, valve guides, and valve seats. Engine coolant inhibits corrosion in the coolant system and lubricates the coolant pump. Windshield washer fluid carries away dirt. Several solid, composite, or polymeric components (e.g., spacers, tensioners, seals, or gasket materials) are also called upon to provide lubrication function. For the substances in a car or light-duty truck that have lubricating properties, a variety of topics are addressed; for example:

• Function of the lubricant

• Range of service conditions that the lubricant must withstand

• Properties and composition of the lubricant

• Changes in lubricant properties during service

• Test methods

• Current issues with regard to the lubricant

6.3 Engine Oil

6.3.1 Composition:

Engine oil is composed primarily of base stock (70 to 90% of the total composition), with the remainder consisting of various protective additives. The base stock can be formed by distilling and collecting a fraction of a batch of crude oil such that the molecular weight of the fraction is within a desired range. This fraction can be purified by solvent extraction and solvent dewaxing so that the composition is appropriate for a lubricant base stock. Base stocks of this type are designated "Group I"

(Note: the American Petroleum Institute [API] has defined a series of base stock groups [currently I through V] according to base oil properties. Oils formulated with the same additive package and base stocks belonging to the same group are considered comparable for the purpose of base oil interchange, and thus may be eligible for a reduced amount of engine testing to certify them for licensing under the terms of the API Engine Oil Licensing and Certification System). If further enhancement of the base stock properties is desired, various chemical reactions can be employed, such as reacting with hydrogen at high temperature and pressure in the presence of a catalyst. This process removes undesirable elements such as nitrogen, oxygen, and sulfur; removes double bonds in the carbon-containing compounds (which creates a structure that is less susceptible to chemical attack); and breaks larger hydrocarbon molecules into smaller, more fluid ones. Further treatment can remove any remaining fractions that are chemically reactive or are likely to solidify at low temperature. The resulting base stock is resistant to chemical attack, fluid at low temperature, and less volatile at higher engine operating temperatures. Fluids processed in some or all of the ways described here are classified as Group II or Group III base stocks. Alternatively, base stock can be formed by chemical reaction of low-molecular-weight building blocks to create a "synthetic" oil whose molecular-weight distribution range is small and whose chemical structure is resistant to degradation. Such oils can be classified as Group IV (for polyalphaolefin base stocks) or Group V base stocks (anything not covered in I through IV). Engine oil base stocks can consist of oils from any combination of these processes. Those base stocks that are derived primarily from the refining process are popularly termed "mineral" oil. Those that are derived primarily from the buildup of low-molecular-weight hydrocarbons are popularly termed "synthetic." "Partial synthetic" and "semi-synthetic" are terms used to describe oils whose base stocks have been enhanced in some manner. In addition, synthetic esters can be added to semi-synthetic and synthetic base stocks to improve seal swell characteristics. Some additive suppliers also find that esters help solubilize additives in synthetic and semi-synthetic base stocks and help keep deposit-forming reaction products in suspension. Among the tribological fluids listed under Lubricant Types, engine oil is relatively complex. As a consequence of the harsh environment that engine oil must withstand, a number of additives are typically incorporated into engine oil to provide specific beneficial properties. One of the most important additives is the antioxidant, which is a sacrificial agent that reduces the tendency of the engine oil to oxidize, thicken, and form varnish, but that eventually becomes inactivated as a consequence of performing its protective function. Anti-wear agents are incorporated to provide a protective, wear-resistant chemical film, especially under high-load conditions. In particular, the zinc dialkyldithiophosphates antioxidant is also an anti-wear agent. Detergents neutralize acids that are formed from combustion or from overheating of engine oil, protect susceptible engine components against corrosion, inhibit deposit formation, and help keep surfaces clean. Dispersants suspend insoluble contaminants and inhibit varnish and sludge formation. Friction modifiers reduce friction, and therefore enhance the energy efficiency of a vehicle. The engine oil's defoamer reduces the tendency of the engine oil to foam, thus ensuring that oil films in critical components will not be weakened by the presence of bubbles. A viscosity modifier or viscosity index improver helps the oil remain fluid when the oil is cold but allows the oil to be sufficiently thick when the oil is hot. Pour-point depressants help keep the oil from solidifying at very low temperature.

6.3.2 Functions of Engine Oil:

Engine oil has many functions. It must withstand the high heat of combustion and must not degrade excessively on exposure to contaminants such as fuel and combustion products. When an engine is newly manufactured, engine surfaces tend to be rougher than is the case after the engine has been in use for a while. During use, these rough surfaces are smoothed, and the metal debris from the smoothing process is transported by the flowing oil past a screen which can remove large particles and debris that is left in the engine from the manufacturing process. Next, the oil passes through a filter that removes smaller particles. The engine oil is then circulated to various engine locations that require specific kinds of protection, as follows. Bearings require sufficient oil pressure that the journal can be supported on an oil film to ensure that contact between journal and bearing is minimized. Engine oil provides a fluid film that lubricates moving valves and carries away heat. Engine oil splashes on the cylinder wall, is carried along the cylinder bore surface by the motion of the piston rings, helps provide a barrier to contain gases formed in the combustion process, and also provides a lubricated surface over which the piston rings can slip easily. Oil jets deliver oil to the under crown of the piston to control temperature and prevent scuffing and ring sticking. Oil splashes or drips on cams and lifters, and causes the formation of a protective chemical film that reduces stress and wear between the cam and lifters. Hydraulic lifters contain oil within their interior to cushion the effect of the motion over the cams and to maintain proper clearance between the valve lifter and camshaft. Oil pressure is used in some applications to adjust the timing chain. Several additional protective capabilities are desirable for engine oil. For example, if oil volatility is low, there will be less tendency of the oil to evaporate on exposure to hot spots, and oil economy may be improved. If oil becomes contaminated with water during short-trip operation, rust-preventatives in the oil will reduce the tendency for the oil to corrode iron in the engine. An increasing concern with modern engine oil is the influence of the engine oil's phosphorus on the deterioration of emission system components (Culley et al., 1995, 1996). Thus, oil formulators are faced with the dilemma of trying to provide adequate wear and antioxidant protection without damaging emissions control devices.

6.3.3 Oil Properties That Are Assessed in Standard Tests:

In many countries, engine oil must pass rigorous standard tests to be certified for use in vehicles. The tests vary from country to country, from one vehicle manufacturer to another, and from one fuel type to another. The standard tests were developed to simulate various types of service that were known to cause oil-related problems. As a consequence of a great amount of effort on the part of those individuals who developed the tests, the standard test methods are designed to ensure that the oil will protect the engine and its components against corrosion and wear; will not thicken excessively during high temperature service; and will withstand the effects of marginal quality fuel in city service in which there is a considerable amount of time spent at idle. There are tests to ensure that the engine oil, as it degrades, will not coat the insides of the engine with sludge or varnish and will continue to protect the engine against excessive rusting, even if water and acids (from the partial combustion of fuel) condense in the engine oil during short-trip service. Engine oil must keep particulates such as soot dispersed well enough that critical flow passages in the engine are not blocked and agglomeration is controlled, so that valve train, ring, and bore wear does not become excessive. The oil must not gel on long-standing, must remain fluid at low temperatures to allow proper engine cranking and oil pumpability, and must remain fuel-efficient during operation.

6.3.4 Effects of Service on Engine Oil Properties:

The tests described in the previous paragraph would not have been developed if the properties of engine oil had remained constant during use. However, engine oil properties change due to such factors as shear of the viscosity modifier or viscosity index improver and fuel entering the oil, which causes lower oil viscosity. Fuel reaction products can condense in the engine oil and inactivate some types of antioxidants (which therefore promotes oil oxidation, formation of acids, and oil thickening). Some fuel reaction products are acids that neutralize the detergent. Nitration of the oil can also occur, via nitrogen in the air or nitrogen reaction products. Such reaction products can cause the formation of a thick black sludge on engine surfaces. Heat from the combustion process lowers oil viscosity, but heat also evaporates the lighter ends of the oil, which eventually causes oil thickening and sluggish flow on cold starts. Causes of changes in oil properties are due to type of service, type of fuel, and characteristics and age of the engine. For example, high-temperature oil degradation is different from low-temperature degradation. In addition, gasoline, diesel oil, alcohol, natural gas, vegetable oil, hydrogen, etc., all cause their own unique type of damage to the oil and to the engine. Service-related changes to the engine oil can typically be categorized under one or more of the following driving styles:

1. Freeway driving

2. High-load or high-speed driving

3. City driving in which the oil is fully warm, but where there are periods at idle

4. Extreme short-trip, cold-start driving in which the engine oil never reaches the equilibrium operating temperature

6.3.4.1 Freeway Service:

In freeway service, the engine oil properties remain stable for a relatively long time. There may be a slight loss of viscosity during the first several thousand kilometers of operation after an oil change as a consequence of shear of the viscosity index improver. Eventually, the oil may thicken from oil oxidation, presence of insoluble contaminants, or evaporation of the lighter ends of the oil. An example of stability of engine oil viscosity during freeway service is shown in Figure 6.1 (Schwartz et al., 1994). At the end of the freeway service, the vehicle began a series of exceedingly short trips in winter. During the short trip service, fuel entered the engine oil and caused a significant drop in viscosity. Figure 6.1 below shows change in engine oil viscosity with distance traveled, freeway service.

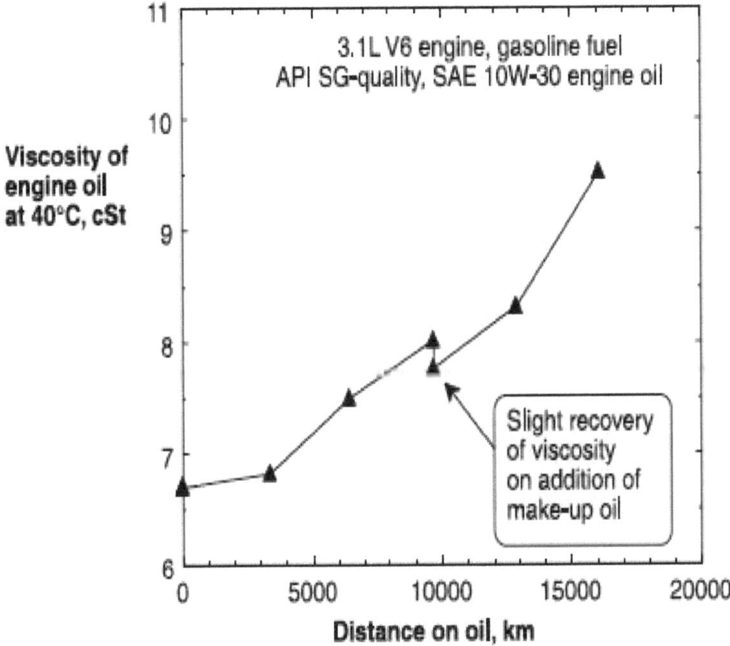

Figure 6.1 Change in Engine Oil Viscosity with Distance Traveled, Freeway Service

6.3.4.2 High Load Service:

In high load service, engine oil tends to thicken as a consequence of either evaporation of the lighter ends of the oil or oxidation of the base stock (or both of these effects simultaneously). Varnish and sludge form and the oil becomes acidic, which puts susceptible engine metals and polymeric materials at risk of corrosion or degradation. Nitration of the oil can cause formation of sludge. Additionally, in diesel engines, soot can enter the engine oil and cause the oil to thicken and to adsorb oil additives. For example, high soot levels can plug filters and cause a high pressure drop across the filter. Dispersants are used in engine oil to counteract some of these adverse soot effects. Soot can contribute to the inactivation of the engine oil's antioxidant. In addition, soot can increase the wear rate by adsorption of antiwear additives and by agglomeration of soot into abrasive particles whose diameter may be greater than the clearances between critical engine components. Figure 6.2 shows an example of the loss of oxidative stability of engine oil in a gasoline-fueled vehicle being driven primarily in freeway service. One of the test vehicles was pulling a 900-kg trailer, and the other test vehicle was not pulling a trailer. Even this relatively minor difference in load can be detected as an increased rate of degradation of the antioxidant. The differences in the rate of loss of oxidative stability are predictable, based on engine measurements such as bulk oil temperature and fraction of the oil volume that is directly exposed to the high heat of the combustion process (Schwartz, 1992a,b). The Sequence IIIE test (ASTM D5533, for oxidation and wear control in oils for gasoline-fueled vehicles); the PSA TU3MH test (CEC L-55-T-95, for ring sticking, piston varnish, and viscosity increase); the Nissan KA24E test (for wear); the L-38 test (ASTM D5119, for high-temperature bearing corrosion); the Toyota IG-FE test (for high-temperature oil oxidation and viscosity increase); and the Mack T-8 test (ASTM D 5967, for ensuring viscosity stability and measuring soot loading in diesel engines) are examples of standard tests that are designed to measure the extent to which an oil's properties remain stable during high-temperature service. (*Note* : Sequence "III" is Roman numeral "three," and "E" represents version "E" of the test method. The next upgrade of the test will be designated Sequence IIIF. Other Sequence tests also use Roman numeral designation: Sequence IID is Sequence 2 version E; Sequence VE is Sequence 5 version E; Sequence VIA is Sequence 6A).

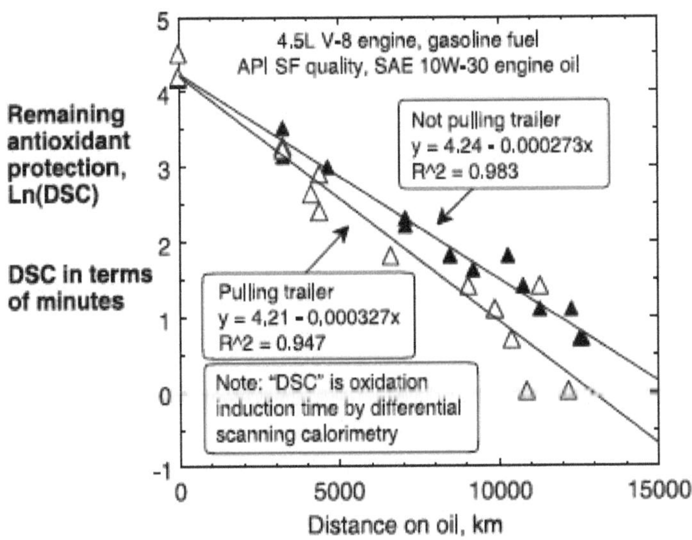

Figure 6.2 Effect of Pulling a 900-kg Trailer on Remaining Antioxidant Protection of Engine Oil, Primarily Freeway Service

6.3.4.3 City Service:

In city service, fuel can accumulate in the engine oil if the engine is at idle for extended periods. Gasoline or diesel fuel in the engine oil can cause the viscosity of the oil to decrease. Incomplete combustion of the fuel can cause the formation of acidic combustion products which condense in the engine oil and begin to inactivate the protective alkaline detergents and dispersants. Acidification of engine oil can put bearings at risk of corrosion. Varnish and sludge can form on engine surfaces. Incomplete combustion can generate soot in diesel engines. An example of the effect of city service on engine oil properties is depicted in Figure 6.3, in which the oil degraded faster in city service than in freeway service. Both vehicles in this test experienced identical driving conditions on a chassis dynamometer. The nature of the fuel can also influence the extent of oil degradation, as suggested by the fact that, in city service, the gasoline-fueled vehicle's viscosity was starting to trend higher than that of the M85-fueled vehicle toward the end of the city-driving test shown in Figure 6.3. The Sequence VE test (ASTM D5302), which measures sludge, varnish, deposits, and valve train wear, is an example of a standard test method that monitors oil life in city service.

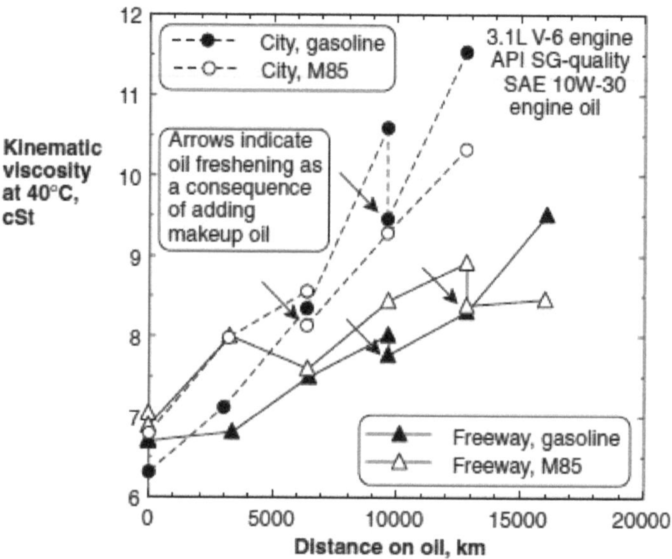

Figure 6.3 Effect of Service Type and Type of Fuel on Engine Oil Viscosity. (Note: M85 means fuel is 85% methanol and 15% gasoline)

6.3.4.4 Extreme Short-Trip, Cold-Start Service:

In extreme short-trip, cold-start driving, the engine oil never warms completely. Fuel, combustion products, and water can accumulate in the engine oil and remain there for a long time (Schwartz, 1991). Engine oil additives can drop to the bottom of the oil pan, carried down by the water that has entered the oil (Schwartz, 1991). Once the engine oil's corrosion inhibitors are inactivated by dropping to the bottom of the pan with the water, the engine is at risk of corrosion. An example of this reversible dropout effect is shown in Figure 6.4 (Younggren and Schwartz, 1993). In a gasoline-fueled vehicle, as a consequence of fuel entering the engine oil, the viscosity of the oil can become considerably lower than one would expect for a given temperature (Schwartz, 1992c). Because the oil temperature is always low in this type of driving, the thickening effect of the engine oil due to low oil temperatures is partially offset by the lower oil viscosity that is a consequence of fuel entering the engine oil during cold starts. Thus, during a cold start, fuel-contaminated engine oil is easier to pump than fresh oil. However, if the driver tows a trailer or takes a long trip after having taken many short trips in winter, the oil's viscosity may be unacceptably low once the oil temperature reaches the normal operating range. Several opposite viscosity effects can also occur. A significant fraction of polar contaminants (derived from partial combustion products of the fuel or aging of the engine oil) in cold engine oil can cause the oil to gel on long standing. And, if the engine is running on an alcohol fuel such as methanol, when alcohol enters the engine oil, it forms an emulsion that can be more viscous than either the methanol or the oil alone. Some of these viscosity effects are shown in Figure 6.5 (Schwartz, 1992c). If an oil-water emulsion is formed, and if this emulsion contaminates a vehicle's positive crankcase ventilation (PCV) emission control system, the PCV control function may not be as effective as it should be. Examples of standard tests that provide information about short-trip, gasoline-fuel service include the ASTM D5844 Sequence IID test and Ball Rust test, both of which assess the ability of the engine to protect engine components against corrosion.

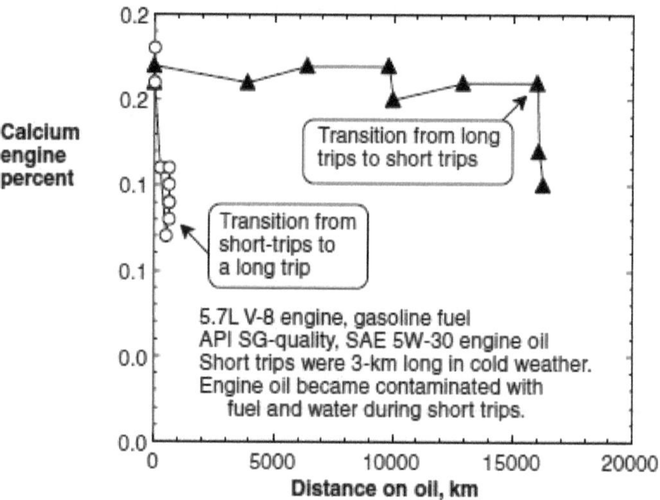

Figure 6.4 Additive Drop Out during Short-Trip Service and Reversibility of the Effect on Commencement of Long-Trip Service

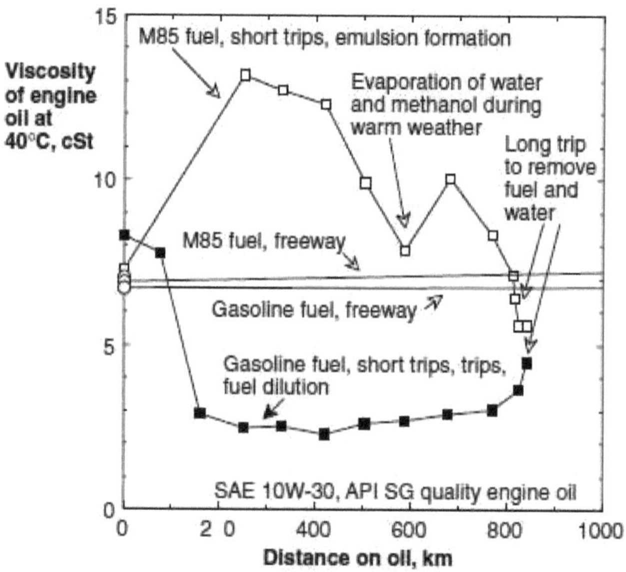

Figure 6.5 Variability of Engine Oil Viscosity with Type of Service and Type of fuel

6.3.5 Fuel Efficiency:

Fuel efficiency in a given engine can be influenced by surface properties of engine components, friction modifying additives in the engine oil, engine oil viscosity, and by changes in these properties with use. Fuel efficiency provided by a given engine oil can be measured by monitoring the amount of fuel consumed in a standard test cycle and comparing the results to the fuel efficiency obtained in an identical test when using a standard engine oil of known fuel efficiency, as described in the Sequence VIA test.

6.3.6 Models of Engine Oil Performance and In Situ Devices to Monitor Oil Properties:

Over the many years since development of the first internal combustion engines, a vast amount of information has been gathered about the lubricant properties required for acceptable engine performance. For example, in recent years, computer models have provided answers to many questions related to engine design and oil performance. Models can predict the desired range for engine oil viscosity, extent of temperature rise in a working bearing, fluid film thickness, and the load that a given material can withstand. As a consequence, great strides have been made with regard to optimizing engine hardware based on computer models. As long as the engine operating conditions are within the range covered by the model, such predictions can usually be trusted. If some failure mode lies outside the domain addressed by the model, it becomes important to define the boundaries of validity of the model, identify the technical disciplines that are pertinent to the failure modes of the problem of interest, and create new classes of models that include all the technical disciplines involved in the problem of interest. As an example of a multidisciplinary approach, a variety of mathematical models are emerging to predict changes in oil properties as a function of engine operating conditions. One of the first of such models was created by Dyson, who considered that an engine could be treated mathematically as if it were a chemical reactor and who predicted the remaining alkalinity in diesel engines by monitoring fuel throughput and sulfur content of the fuel (Dyson et al., 1957). Schwartz found that the remaining useful life of the engine oil's antioxidant in gasoline-fueled engines could be modeled by assuming that the antioxidant was completely inactivated in the small volume of oil that was directly exposed to the high heat of combustion at the top ring reversal point, top dead center (Schwartz, 1992a,b). Audette and Wong were able to estimate the oil film thickness, changes in the oil composition, and the amount of oil consumption that occurred as a consequence of evaporation of the lighter ends of the engine oil from the cylinder bore surface in a diesel engine as a function of the boiling point range of the engine oil, the temperature of the cylinder liner, the volume of oil that was carried to the upper cylinder area, and the characteristics of the piston rings (Audette et al., 1999). For fuels containing both methanol and gasoline, in short-trip service in which both fuel and water enter the engine oil, Schwartz found that emulsion-forming contaminants (methanol, water) in the engine oil caused a viscosity increase, solution-forming contaminants (gasoline) caused a viscosity decrease, and these combined effects on viscosity could be predicted based on the volume fraction of each type of contaminant in the engine oil (Schwartz, 1992c). An area of active investigation is the development of models to predict the end of useful life of engine oil.

Because the nature of oil aging depends on engine design and type of service, incorporating these parameters into models of engine oil life will provide challenging opportunities for future model makers. An additional area that is sure to expand is the development of sensors that monitor one or more properties of engine oil during service. For example, oil viscosity sensors and conductance measurement devices are being developed, patented, and in some cases placed in production (Hell wig et al., 1998; Johnson et al., 1994; Kato et al., 1986; Kauffman, 1989; Kollmann et al., 1998; Wang et al., 1994). As witnessed by the diversity and success of the few examples described here, the process of developing sensors and models of oil performance that comprehend both changing oil properties and service conditions has already begun.

6.4 Automatic Transmission Fluid

Automatic transmission fluids assist the transmission to shift gears. Under high load conditions, the automatic transmission fluid must be able to withstand the heat and forces that are generated. However, unlike engine oil, the transmission fluid is not exposed to contamination from the combustion process. Thus, transmission fluid does not experience the type of degradation modes that occur as a consequence of fuel and its combustion products entering the engine oil. Automatic transmission fluids must resist chemical changes (such as oil thickening and acidification) that are a consequence of oxidation and must minimize the adverse consequences of physical changes such as the low viscosity of hot oil and the high viscosity of cold oil. That is, transmission fluids must retain the ability to flow into critical areas when cold, but not flow out excessively when the oil is hot. The fluids need to maintain the appropriate amount of friction and reduce wear and corrosion, either by providing a fluid film between moving components or providing a protective chemical film on the surface. Transmission fluids need to maintain the appropriate amount of seal swell to reduce leakage, as well as retain stable shift properties, even after extended service. The fluids carry away heat and debris. They must also retain friction modifier durability in the harsh environment of constant slipping in electronically controlled slipping torque-converter clutches.

6.4.1 Composition:

All automatic transmission fluids recommended for use by car manufacturers are formulated using synthetic or mineral oil base stock plus a variety of additives. Synthetic fluids tend to provide properties that are more stable and last longer than mineral oil, but they also cost more. Additives in the oil include friction modifiers to produce the desired clutch operation, corrosion inhibitors, viscosity index improvers, dispersants, seal swell agents, antioxidants and antiwear agents, with a total additive treatment level typically in the range of 10 to 20% (Kemp et al., 1990). Among these additives, friction modifiers are of particular importance for proper transmission performance. The chemicals employed to produce good clutch friction characteristics are often long-chain hydrocarbon molecules with a polar group on one end. An example is dilauryl phosphate, which is composed of two chains, each containing 12 carbon atoms, with a phosphorus-containing acid polar group on one end of the dual chain. Another example is oleic acid, which is an 18-carbon chain with an organic acid polar group on the end. Both dilauryl phosphate and oleic acid are soluble in mineral oil, and their effectiveness in modifying clutch friction characteristics depends on the capability of the polar group to attach to the rubbing surfaces. Friction-modifier additives prevent surface adhesion and surface asperity contact that would generate high levels of friction and wear. The polar groups of friction modifiers either react chemically with the rubbing surfaces or physically adsorb onto the surfaces to form a tenacious film. Clutch friction characteristics using three friction-reducing additives (oleic acid, dilauryl phosphate, and glycol dioleate) formulated with mineral base stock have been compared in a laboratory clutch friction device (Tung et al., 1989; Stebar et al., 1990). Among these three additives, dilauryl phosphate demonstrates the most desirable clutch friction characteristics.

6.4.2 Evaluating Automatic Transmission Fluids:

To ensure that the properties of automatic transmission fluids remain within an acceptable range, the fluids must meet various chemical and physical requirements. The fluids are also evaluated using performance tests that measure such characteristics as wear resistance, retention of appropriate friction properties, torque, time-to-lock-up, and shift-feel. References to North American test methods are provided in the *1999 SAE Handbook*, Volume 1, "Fluid for Passenger Car Type Automatic Transmissions SAE J311 Feb 94 SAE Information Report,", and in the requirements of the various automobile manufacturers.

6.4.2.1 Physical and Chemical Tests:

The following tests are used to assess the physical and chemical characteristics of automatic transmission fluids:

1. Flash point and pour point
2. Fluid oxidation

3. Kinematic viscosity at 40°C and 100°C

4. Brookfield viscosity at various temperatures, including tests down to –40°C

5. Copper corrosion

6. Rust prevention

7. Foaming tendency

8. Elastomer compatibility

6.4.2.2 Transmission Fluid Oxidation Tests:

In the Transmission Fluid Oxidation test (a General Motors DEXRON®-III specification) formerly known as THOT (Turbo Hydramatic Oxidation Test), the transmission must provide satisfactory operation at an elevated temperature (163°C) for 300 hours in a full-scale transmission test. Air is passed through the transmission while the transmission is being motored. At the end of the test, the transmission fluid is evaluated for such oil properties as acidity, viscosity, and corrosion resistance. The transmission is also evaluated by comparing the parts to those obtained in a test using a reference oil. The Aluminum Beaker Oxidation Test (ABOT) is a bench oxidation test (used in the Ford Motor Company MERCON® specification) that also measures the ability of automatic transmission fluid to resist oxidization.

6.4.2.3 Transmission Cycling Test:

The Transmission Cycling Test (DEXRON-III specification) uses a V8 engine to drive a transmission. The transmission goes through 20,000 cycles of a test that includes acceleration from first through fourth gear. Each cycle lasts approximately 45 seconds, so that the test time is approximately 220 hours. Fluid temperature is maintained at an elevated temperature, and shift characteristics are monitored throughout the test. At the end of the test, the chemical properties of the fluid and the shift times must remain within specified limits, and the condition of the transmission parts must be equal to or better than those tested using a standard reference oil.

6.4.2.4 Plate Friction and Band Friction Tests:

An SAE No. 2 Machine (electrically driven inertial device) is used to evaluate the frictional characteristics of transmission fluids under both static and dynamic conditions. Dynamic friction levels are evaluated by engaging the clutch plates immediately after the electric motor is shut off, using the clutch plate friction to stop the rotating inertia of the electric motor and flywheel. The change of friction after thousands of lock-up cycles is evaluated as a measure of fluid durability. In a similar test, band friction is evaluated. Test requirements include no unusual wear or flaking on test parts and limits on changes in torque and stop time.

6.4.2.5 Shudder and Stick-Slip Test:

Shudder is low-frequency vibration that occurs mainly between 20 and 30 Hz. It is a resonance phenomenon that is influenced by both the stick-slip friction characteristics of the automatic transmission fluid and the materials in the transmission. The relationship between the coefficient of friction and the sliding speed is shown in Figure 6.6. The condition in which the friction coefficient decreases at decreasing sliding speed reduces the tendency for stick-slip, thus decreasing the probability of shudder. The condition of low friction at high speed, and high friction at low speed, fosters the generation of stick-slip or shudder (Ward et al., 1994). Mechanical design and transmission control can influence the susceptibility of a transmission to shudder. "Green shudder" can occur in a new transmission. Figure 6.7 is a diagram of a low-velocity friction apparatus for determination of the low-speed friction coefficient as a function of sliding speed. This type of testing is used to study the stick-slip durability of automatic transmission fluids with different friction materials. Variable-speed friction testers also provide valuable information.

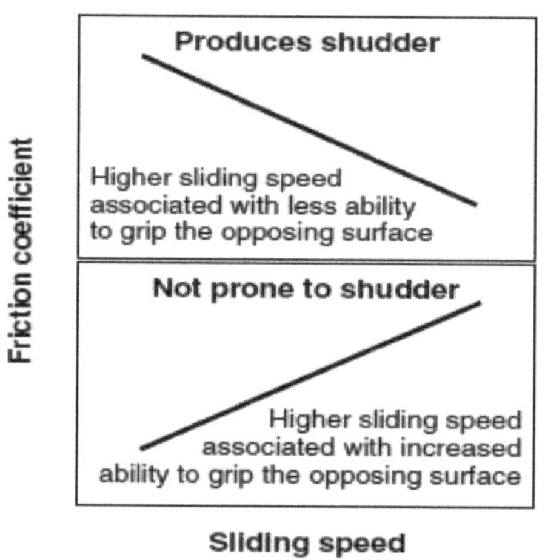

Figure 6.6 Relationship between Sliding Speed and Clutch Frictional Characteristics

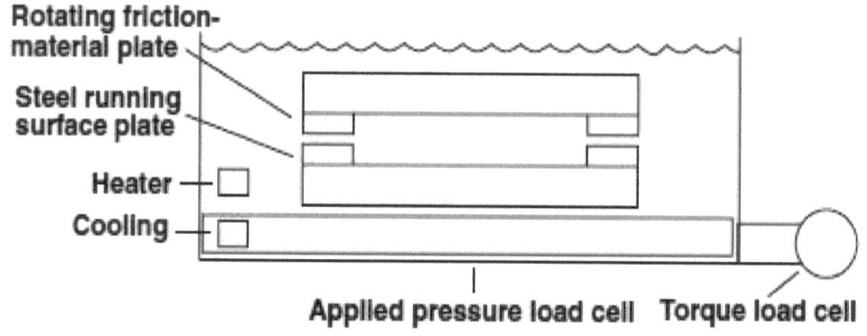

Figure 6.7 Test Apparatus to Measure Friction Characteristics at Low Sliding Speed

6.4.2.6 Automatic Transmission Fluid Road Testing:

To determine shift-feel characteristics, test oils are evaluated in road tests and compared to various reference oils in specific transmission and vehicle combinations (Ashikawa et al., 1993).

6.4.3 Current Automatic Transmission Fluid Specifications:

Automatic transmission fluid specifications for commercially approved fluids have been significantly revised since approximately 1990. As a consequence of the development of continuous-slip torque converter clutches to improve fuel economy, automatic transmission fluid specifications have more stringent friction performance requirements than formerly. Two commercial specifications provide the primary transmission fluid performance requirements for North American cars and light-duty trucks: the DEXRON-III specification for General Motors automatic transmissions and the MERCON specification for Ford Motor Company automatic transmissions. Typical test requirements include most of the following: Vickers vane pump wear test, various bench tests, plate clutch friction test, band clutch friction test, oxidation test, cycling test, vehicle shift evaluation, electronic controlled converter clutch vehicle test, sprag clutch wear test, and the aluminum beaker oxidation test.

6.5 Lubricants

6.5.1 Air-Conditioning Lubricant:

Air-conditioning lubricants are fluids that lubricate the moving components of an automotive air conditioning system. The lubricant is transported through the air-conditioning system by the air conditioning fluid, and therefore an air conditioning lubricant must lubricate even when diluted with a significant amount of the air-conditioning medium. In addition, the lubricant must be sufficiently compatible with the air-conditioning fluid so that the fluid will carry the lubricant to all the critical parts of the air-conditioning system. Prior to 1995, R12 (which contains chlorine) was a major automotive air-conditioning fluid, and mineral oil was its associated lubricant. As a consequence of political and environmental concerns (e.g., the formation of the hole in the protective ozone layer in the upper atmosphere), the use of R12 was restricted by international agreement, and a search began for an alternative air-conditioning fluid. R134a (which contains fluorine rather than chlorine) was identified as a candidate replacement for R12. However, it soon became evident that mineral oil was not an optimum lubricant for use with R134a because mineral oil was not sufficiently transported through the air conditioner by R134a. Thus, it became necessary to identify a lubricant that could be transported by R134a and could protect the moving components of the air-conditioning system. After testing a variety of candidates, it was found that oxygen-containing lubricants such as polyglycols could be transported by R134a. This development process highlights the fact that if any component of a lubrication system (fluid, materials, design) is changed, tests must be conducted to determine whether any incompatibilities have arisen.

6.5.2 Gear and Axle Lubricants:

Gear lubricants are used to lubricate manual transmissions and axles. Gear lubricants must contain appropriate additives and must exhibit some or all of the following properties, depending on the application: thermal and oxidative stability, resistance to extreme pressure, and protection against corrosion, wear, and spalling. It has been shown that, through proper selection of gear lubricants, vehicle fuel economy can be improved. Gear lubricants must provide seal protection and appropriate frictional behavior. Gear and axle lubricants promote temperature reduction via friction reduction. In addition, they must maintain appropriate viscosity at both high and low temperatures, and must remain stable over long periods of time. Test methods for gear lubricants include such operating conditions as high torque at low speed and high speed with shock loading. Lubricant test measurements include oxidation resistance, protection against wear and material failure, corrosion protection, retention of desired friction characteristics, acceptable viscosity at both low and high temperatures, and seal compatibility. Gear lubricants used in manual transmissions must have appropriate friction characteristics to ensure proper functioning of the transmission synchronizers to prevent gear clashing and allow smooth shifting.

6.5.3 Grease:

Grease is a solid or semi-solid lubricant that is used whenever a lubricant must remain in the place in which it is needed. Grease is composed of oil or fluid, thickener, and additives. A typical composition would be petroleum oil thickened with soap, and additives as needed to impart the desired chemical and physical properties. Grease is used in more locations in a vehicle than any other lubricant, including such diverse places as door locks, remote-control mirror assembly, seat adjustment gears, window levers, windshield wiper gear, water-pump bearing, horn and other electrical contacts, brake system, heating and cooling system, wheel bearings, U-joint, and many other places in the vehicle's powertrain. A good grease will resist oxidation and degradation, provide an anti-wear fluid film, and prevent metal to metal contact. Greases must often provide various types of chemical protection such as extreme pressure agents, and corrosion and rust inhibition. Greases are often called upon to keep contaminants away, assist in the sealing process, and resist being washed or squeezed out of their desired location. The base stock in grease makes up 70 to 90+% of the grease composition and may be composed of either synthetic or mineral oil. Additives and thickeners represent the remainder of the composition. Tests for grease typically include measuring the hardness or consistency, and high and low temperature properties. Additional tests include corrosion resistance, stability in the presence of water, oxidative and shear stability, load-carrying capacity, wear resistance, and life performance properties.

6.6 Fluids

6.6.1 Power Steering Fluid:

Some power steering systems may require a special power steering fluid; others may use one of the lubricants designed for some other application, such as transmission fluid or engine oil. Those fluids that are specialized for power steering applications must resist wear, rust, and oxidation, while maintaining the appropriate viscosity. Those fluids that are specialized for other applications typically do not need to meet any additional requirements when used as a power steering fluid.

6.6.2 Brake Fluid:

Brake fluids typically are either silicone based or glycol based. They must remain relatively stable during use, maintain appropriate alkalinity to resist becoming corrosive, remain fluid at low temperature, not cause elastomers to deteriorate, and be water tolerant. Because brake fluid can retain water, it needs to be changed once the water content becomes too high for proper braking performance.

6.6.3 Shock Absorber Fluid:

Shock absorber fluid is typically a hydraulic fluid that cushions the shock when the wheels of a vehicle travel over a rough surface. Shock absorber fluid needs to remain stable during use, not entrain an excessive amount of air, exhibit good sealing performance, and have the appropriate viscosity characteristics.

6.6.4 Fuel:

Various types of fuels are used in internal combustion engines. Gasoline, diesel oil, compressed natural gas, ethanol, and methanol are some that have been used in recent years. In a development program in which an existing engine is converted for use with a fuel that is different from the one for which the engine was originally designed, one must pay particular attention to material compatibility. The fuel must not cause unacceptable wear or deposits in injectors, fuel pump, or other components in the fuel system, and must not corrode the metals that it touches. In addition, the fuel should not generate so much soot that the engine oil becomes unacceptably contaminated. The fuel must be compatible with the materials in the fuel system, and must also be compatible with the materials in the lubrication system (because under some driving conditions, fuel condenses in the engine oil). If an alternative fuel is used, seals and polymeric or composite engine components must be tested for compatibility, because fuel that has accumulated in the engine oil can extract beneficial plasticizers from elastomeric materials, causing the elastomers to shrink, leak, or harden. If a fuel has a composition that is soluble in a given elastomer, the fuel may cause the elastomer to swell excessively. An example of the effects of an alternative fuel (M85: 85% methanol, 15% gasoline) on several engine elastomers is shown in Figure 6.8 (Schwartz, 1988). Alternative fuels that form emulsions in engine oil (rather than true solutions) can cause engine oil to thicken — as opposed to the effect of gasoline in oil, which causes oil to become less viscous. Changes in viscosity due to fuel and water in engine oil were described in chapter 6 and shown in Figure 6.5.

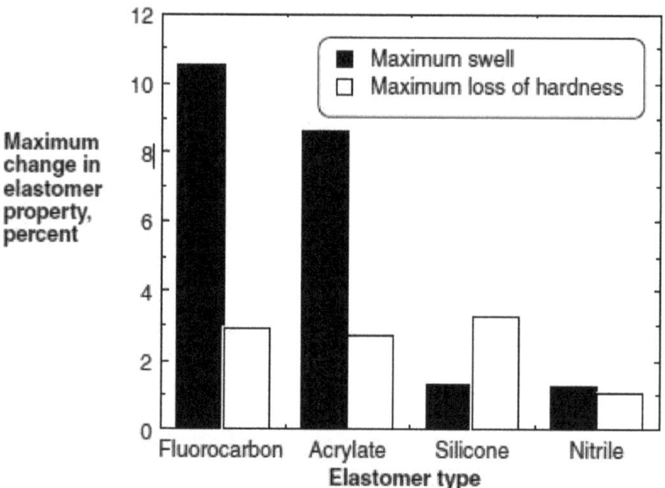

Figure 6.8 Effect of an Alternative Fuel (M85) on Seal Properties

6.6.5 Functional Fluids:

Engine Coolant, Windshield Washer Fluid:

Various fluids within a vehicle have some properties that are related to lubrication. For example, the engine coolant/antifreeze must carry away heat, must inhibit wear in the pump, and must not corrode materials or damage seals in the engine coolant system. Engine coolant is typically composed of water and glycol in such proportion that the freezing point is well below that of water and the boiling point is significantly above that of water. Corrosion inhibitors in the coolant must protect all the metals found in the coolant system. Windshield washer fluid should remain fluid at low temperatures, carry away dirt, and should not damage paint, windshield-wiper blades, or other materials.

Chapter Seven

Conclusions

Fluids and components that provide a lubricating function in a vehicle differ widely. For example, engine oil must remain reasonably stable even if it becomes contaminated with fuel and combustion products. Transmission fluid must maintain anti-wear, oxidation, and friction modifier functionality under extremely harsh temperature conditions. Many different kinds of greases are required to satisfy a wide variety of conditions in automotive applications. To provide durable, problem-free vehicle performance, an awareness of lubricant properties and the ways in which they change during service will assist those who design components, assess component durability, choose materials, and model lubricant performance. The major issues on which automotive lubricants have a significant impact include energy conservation and reduction of pollution. Research is in progress on both of these issues. In addition, vehicle manufacturers are investigating all possible ways in which vehicle weight might be reduced and ways in which manufacturing might be made more efficient (via improved manufacturing procedures, recycling, scrap reduction, etc.). All these issues must be addressed without compromising the safety of those operating the vehicles and those working in the manufacturing plants.

References

Anderson, A.E. (1992), Friction and wear of automotive brakes, in ASM Handbook, Vol. 18, Blau, P.J. (Ed.), 569-577.

Andersson, B.S. (1991), Company Perspectives in Vehicle Tribology — Volvo, in 17th Leeds-Lyon Symposium on Tribology — Vehicle Tribology, Tribology Series, Vol. 18, Elsevier Science, Oxford, U.K., 503-506.

Archard, John Frederick (1 August 1953). "Contact and Rubbing of Flat Surfaces". Journal of Applied Physics. 24(8): 981–988. Bibcode: 1953JAP....24...981A. Doi: 10.1063/1.1721448. ISSN 0021-8979.

Ashikawa, R., Naruse, T., Kuroshima, H., Matsuoka, T., Adachi, T., and Nakayama, T. (1993), ATF Characteristics Required for the Latest Automatic Transmissions, SAE Paper No. 932849.

Audette III, W.E. and Wong, V.W. (1999), A Model for Estimating Oil Vaporization from the Cylinder Liner as a Contributing Mechanism to Engine Oil Consumption, SAE Paper No. 1999-01-1520.

Bell, J.C. (1998), Gasoline engine valve train design evolution and the antiwear requirements of motor oils, J. Engine Tribol., Proc. Inst. Mech. Engrs., 212(J4), 243-257.

Booker, J.F. (1965), dynamically loaded journal bearings: mobility method of solution, Trans. ASME, J. Basic Eng., D, 187, 537-546.

Booker, J.F. (1969), dynamically loaded journal bearings: maximum film pressure, Trans. ASME, J. Lubr. Tech., July, 534-543.

Bowden, Frank Philip; Tabor, David (2001). The Friction and Lubrication of Solids. Oxford Classic Texts in the Physical Sciences. ISBN 9780198507772.

Brauers, B. (1988), NF Steel Rail Oil Control Ring — Test Results and Further Development, Goetze AG, Technical Paper K42.

Bush, G.P., Fox, M.F., Picken, D.J., and Butcher, L.F. (1991), Composition of lubricating oil in the upper ring zone of an internal combustion engine, in Tribol. Int., 24(4), 231-233.

Buuck, B.A. (1982), Elementary Design Considerations for Valve Gears, SAE Paper No. 821574.

Chaston, J.C. (1 December 1974). "Wear resistance of gold alloys for coinage". Gold Bulletin. 7 (4): 108–112. Doi: 10.1007/BF03215051. ISSN 0017-1557.

Clark, S.E. (1971), the contact between tire and roadway, in Mechanics of Pneumatic Tires, Clark, S.E. (Ed.), National Bureau of Standards Monograph 122.

Coy, R.C. (1997), Practical applications of lubrication models in engines, New Directions in Tribology, MEP, 197-209.

Culley, S.A. and McDonnell, T.F. (1995), the Impact of Passenger Car Motor Oil Phosphorus Levels on Engine Durability, Oil Degradation, and Exhaust Emissions in a Field Trial, SAE Paper No. 952344.

Culley, S.A., McDonnell, T.F., Ball, D.J., Kirby, C.W., and Hawes, S.W. (1996), the Impact of Passenger Car Motor Oil Phosphorus Levels on Automotive Emissions Control Systems, SAE Paper No. 961898.

Dowson, D. and Higginson, E. (1991), Elasto-Hydrodynamic Lubrication, Pergamon Press.

Dowson, Duncan (1997). History of Tribology (Second edition). Professional Engineering Publishing. ISBN 1-86058-070-X.

Dyson, A., Richards, and L.J., and Williams, K.R. (1957), Diesel engine lubricants: their selection and utilization with particular reference to oil alkalinity, in Proc. Inst. Mech. Eng. 171, 717-740.

Fessler, R. (1999), U.S. Department of Energy Workshop on Industrial Research Needs for Reducing Friction and Wear, Argonne National Laboratory.

Field, J. (2008). "David Tabor. 23 October 1913 – 26 November 2005". Biographical Memoirs of Fellows of the Royal Society. 54: 425–459. Doi: 10.1098/rsbm.2007.0031.

French, T. (1988), Tyre Technology, Adam Hilger.

Gangopadhyay, A., McWatt, D., Willermet, P., Crosbie, G.M., and Allor, R.L. (1999), Effects of composition and surface finish of silicon nitride tappet inserts on valve train friction, in Lubrication at the Frontier, Leeds-Lyon Symposium on Tribology, Lyon, France, Tribology Series Vol. 36, Elsevier, p. 891.

Gao, Jianping; Luedtke, W. D.; Gourdon, D.; Ruths, M.; Israelachvili, J. N.; Landman, Uzi (1 March 2004). "Frictional Forces and Amontons' Law: From the Molecular to the Macroscopic Scale". The Journal of Physical Chemistry B. 108 (11): 3410–3425. Doi:10.1021/jp0363621. ISSN 1520-6106.

Hata, H. and Tsubouchi, T. (1998), Molecular structure of traction fluids in relation to traction properties, Tribol. Lett. 5, 69-74.

Heilich, F. (1983), Traction Drives: Selection and Application, Marcel Dekker, 18-51.

Heizler, H. (1999), Vehicle and Engine Technology, 2nd ed., Arnold, London, 783.

Hellwig, G., Normann, N., and Uhl. G. (1988), Ein Sensor auf dielektrischer Basis zur On-Line-Charakterisierung von Motorenölen (Alkalinität, Viskosität), Mineralöl. Techn. 10, 1-26.

Hendriks, E. (1993), Qualitative and Quantitative Influence of Fully Electronically Controlled CVT on Fuel Economy and Vehicle performance, SAE Paper No. 930668.

Hendriks, E. (1995), Second generation technology repositions CVT, in VDI Berichte, 1175, 603-619.

Hewko, L.O., Rounds, F.G., and Scott, R.L. (1962), Traction Capacity and Efficiency of Rolling Contacts, Elsevier Publishing Corp.

Hsu, S. (1995), National Institute of Science and Technology (NIST) Engine Materials and Tribology Workshop, Gaithersburg, MD, 3-5.

Hutchings, Ian M. (15 August 2016). "Leonardo da Vinci's studies of friction" (PDF). Wear. 360(Supplement C): 51–66. Doi: 10.1016/j.wear.2016.04.019.

Johnson, M.D., Korcek, S., and Schriewer, D. (1994), In-Service Engine Oil Condition Monitoring Opportunities and Challenges, SAE Paper No. 942028.

Jost, Peter (1966). "Lubrication (Tribology) - A report on the present position and industry's needs". Department of Education and Science. London, UK: H. M. Stationery Office.

Kato, T. and Kawamura, M. (1986), Oil maintenance tester: a new device to detect the degradation level of oils, Lubr. Eng., 42(11), 694-699.

Kauffman, R.E. (1989), Development of a remaining useful life of lubricant evaluation technique. III. Cyclic voltammetric methods, Lubr. Eng., 45(11), 709-716.

Kemp, S.P. and Linden, J.L. (1990), Physical and Chemical Properties of a Typical Automatic Transmission Fluid, SAE Paper No. 902148.

Kluger, M. and Fussner, D.R. (1997), An Overview of Current CVT Mechanisms, Forces, and Efficiencies, SAE Paper No. 970688.

Kollmann, K., Gürtler, T., Land, K., Warnecke, W., and Müller, H.D. (1998), Extended Oil Drain Intervals Conservation of Resources or Reduction of Engine Life (Part. II), SAE Paper No. 981443.

Korcek, S., Jensen, R.K., Johnson, M.D., and Sorab, J. (1999), Fuel efficient engine oils, additive interactions, boundary friction and wear, in Lubrication at the Frontier, Proc. Leeds-Lyon Symp. On Tribology, September 1998, Elsevier, Tribology Series 36, pp. 13-24.

Li, D.F., Rohde, S.M., and Ezzat, H.A. (1982), an automotive piston lubrication model, ASLE Trans., 26(2), 151-160.

Machida, H. and Kurachi, N. (1990), Prototype Design and Testing of a Half Toroidal CVT, SAE Paper No. 900552.

Martin, F.A. (1983), Developments in Engine Bearings, in Tribology of Reciprocating Engines, Proc. 9th Leeds-Lyon Symp. Tribology, 1982, Leeds, Butterworth, 9-28.

Massey, I.D., MacQuarrie, N.A., Coston, N.F., and Eastham, D.R. (1991), Development of crankshaft bearing materials for highly loaded applications, in Vehicle Tribology, Proc. 17th Leeds-Lyon Symp. Tribology, September 1990, Leeds, Tribology Series, Vol. 18, Elsevier, Amsterdam, 43-52.

Mitchell, Luke (November 2012). Ward, Jacob (edition). "The Fiction of Nonfiction". Popular Science. No. 5. 281(November 2012): 40.

Monaghan, M.L. (1987), Engine Friction a Change in Emphasis, Inst. Mech. Engrs., 2nd BP Tribology Lecture.

Monaghan, M.L. (1989), Putting friction in its place, 2nd Int. Conf.: Combustion Engines Reduction of Friction and Wear, in Inst. Mech. Eng. conf. pub. 1989-9, Paper C375/KN1, 1-5.

Moore, D.F. (1975), Principles and Applications of Tribology, Pergamon Press.

Munro, R. (1990), Emissions Impossible — the Piston and Ring Support System, SAE Paper No. 900590.

Nakasa, M. (1995), Engine Friction Overview, in Proc. Int. Tribol. Conf., Yokohama, Japan.

Narasimhan, S.L. and Larson, J.M. (1985), Valve Gear Wear and Materials, SAE Paper No. 851497.

Neale, M.J. (1994), Drives and Seals, a Tribology Handbook, 1st ed., Butterworth-Heinemann, Oxford, U.K.

Neale, Michael J. (1995). The Tribology Handbook (2nd Edition). Elsevier. ISBN 9780750611985.

Nunney, M.J. (1998), Automotive Technology, SAE International, Butterworth-Heinemann.

Oh, K.P., Li, C.H., and Goenka, P.K. (1987), Elastohydrodynamic lubrication of piston skirts, J. Tribol. Trans. ASME, 109, 356-362.

Pillai, P.S. (1992), Friction and wear of tires, in ASM Handbook, Vol. 18, Friction, Lubrication and Wear Technology, Blau, P.J. (Ed.), ASM International, 578-581.

Popova, Elena; Popov, Valentin L. (1 June 2015). "The research works of Coulomb and Amontons and generalized laws of friction". Friction. 3 (2): 183–190. Doi: 10.1007/s40544-015-0074-6.

Pottinger, M.G. and Yager, T.J. (1986), The tyre pavement interface, in STP 929, American Society for Testing and Materials, IX-XV.

Priest, M., Dowson, D., and Taylor, C.M. (1999), Predictive wear modeling of lubricated piston rings in a diesel engine, Wear, 231(1), 89-101.

Priest, M. and Taylor, C.M. (1998), Automobile engine tribology approaching the surface, in AUSTRIB '98: Tribology at Work, The Institution of Engineers, Australia, 353-363.

Reye, Karl Theodor (1860) [1859-11-08]. Bornemann, K. R. (edition). "Zur Theorie der Zapfenreibung" [On the theory of pivot friction]. Der Civilingenieur - Zeitschrift für das Ingenieurwesen. Neue Folge (NF) (in German). 6: 235–255. Retrieved 25 May 2018.

Ruddy, B.L., Dowson, D., and Economou, P.N. (1982), a review of studies of piston ring lubrication, in Proc. 9th Leeds-Lyon Symp. On Tribology: Tribology of Reciprocating Engines, Paper V (i), 109-121.

Rühlmann, Moritz (1885). Vorträge über die Geschichte der technischen Mechanik und theoretischen Maschinenlehre und der damit im Zusammenhang stehenden mathematischen Wissenschaften. Teil 1. Georg Olms Verlag. p. 535. ISBN 978-3-48741119-4 – via Google Books.

Rycroft, J.E., Taylor, R.I., and Scales, L.E. (1997), Elastohydrodynamic effects in piston ring lubrication in modern gasoline and diesel engines, in Elastohydrodynamic Fundamentals and Applications in Lubrication and

Traction, Proc. 23rd Leeds-Lyon Symposium on Tribology, September 1996, Dowson, D. et al. (Eds.), Elsevier, pp. 49-54.

Schwartz, S.E. (1988), Effects of methanol, water, and engine oil on engine lubrication-system elastomers, Lubr. Eng., 44(3), 201-205.

Schwartz, S.E. (1991), Observations through a Transparent Oil Pan during Cold-Start, Short-Trip Service, SAE Paper No. 912387.

Schwartz, S.E. (1992a), a model for the loss of oxidative stability of engine oil during long-trip service. I. Theoretical considerations, STLE Tribol. Trans., 35(2), 235-244. Schwartz, S.E. (1992b), a model for the loss of oxidative stability of engine oil during long-trip service. II. Vehicle measurements, STLE Tribol. Trans., 35(2), 307-315. Schwartz, S.E. (1992c), a Comparison of Engine Oil Viscosity, Emulsion Formation, and Chemical Changes for M85 and Gasoline-Fueled Vehicles in Short-Trip Service, SAE Paper No. 922297.

Schwartz, S.E. and Mettrick, C.J. (1994), Mechanisms of Engine Wear and Engine Oil Degradation in Vehicles Using M85 or Gasoline, SAE Paper No. 942027.

Stebar, R.F., Davison, E.D., and Linden, J.L. (1990), Determining Frictional Performance of Automatic Transmission Fluids in a Band Clutch, SAE Paper No. 902146.

Tabor, D. (1 November 1969). "Frank Philip Bowden, 1903–1968". Biographical Memoirs of Fellows of the Royal Society. 15 (53): 317. Bibcode: 1969JGlac...8...317T. Doi: 10.1098/rsbm.1969.0001. ISSN 0080-4606.

Taylor, C.M. (Ed.) (1993), Engine tribology, in Tribology Series, Vol. 26, Elsevier, Amsterdam, 301.

Taylor, C.M. (1994), Fluid film lubrication in automobile valve trains, J. Eng. Tribology, Proc. Inst, Mech. Engs., 208(J4), 221-234.

Taylor, C.M. (1998), Automobile engine tribology design considerations for efficiency and durability, Wear, 221(1), 1-8.

Tsubouchi, T., Hata, H., Yamada, H., and Aoyama, S. (1990), Development study of new traction fluids for automotive use, in Proc. 17th Leeds-Lyon Symp., 433-443.

Tung, S.C. and Hartfield-Wünsch, S. (1995a), Advanced engine materials: current development and future trends, in NIST (National Inst. of Science and Technology), Engine Materials and Tribology Workshop, Gaithersburg, MD, April 4-7, 1995.

Tung, S.C. and Hartfield-Wünsch, S. (1995b), the effect of microstructure on the wear behavior of thermal spray coatings, in ASM Trans., 107.

Tung, S.C., Hill, S., and Hartfield-Wünsch, S. (1996), Bench wear testing of common gasoline engine cylinder bore surface/piston ring combinations, STLE Tribol. Trans., 39(4), 929-935.

Tung, S.C. and Wang, S. (1989), Using electrochemical and spectroscopic techniques as probes for investigating metal-lubricant interactions, STLE Tribol. Trans., 33(4), 563-572.

Veith, A.G. (1986), The tyre pavement interface, in STP 929, American Society for Testing and Materials, 125-158.

Wang, S.S., Lee, H.S., and Smolenski, D.J. (1994), the development of in situ electrochemical oil condition sensors, Sensors and Actuators: B. Chemical, 17(3), 179-185.

Ward, W.C. and Copes, R.G. (1993), the next ATF challenge — meeting fuel efficiency needs and providing driver satisfaction, in Proc. NPRA Annual Meeting, AM-93-06.

Ward, W.C., Sumiejski, T.A., and Schiferi, E.L. (1994), Friction and Stick-Slip Durability Testing of ATF, SAE Paper No. 941883.

Watts, R.F. et al. (1990), Prediction of low speed clutch shudder in automatic transmissions using the low velocity friction apparatus, in 7th Int. Colloq. on Automotive Lubrication, Technische Academie, Esslingen, Germany.

Watts, R.F. and Richard, K.M. (1997), CVT lubrication: the effects of lubricants on the frictional characteristics of the belt-pulley interface, in 5th CEC Int. Symp. Performance Evaluation of Automotive Fuels and Lubricants, Sweden.

Wills, G.J. (1980), in Lubrication Fundamentals, Marcel Dekker, Inc.

Xu, H. (1999), recent advances in engine bearing design analysis, J. Eng. Tribol., Proc. Instn. Mech. Engrs., 213(J4), 239-251.

Younggren, P.J. and Schwartz, S.E. (1993), The Effects of Trip Length and Oil Type (Synthetic vs. Mineral Oil) on Engine Damage and Engine-Oil Degradation in a Driving Test of a Vehicle with a 5.7L Engine, SAE Paper No. 932838.

Author Brief Auto - Biography

Dr. Engineer **Osama Mohammed Elmardi Suleiman Khayal** was born in Atbara, Sudan in 1966. He received his diploma degree in mechanical engineering from Mechanical Engineering College, Atbara, Sudan in 1990. He also received a bachelor degree in mechanical engineering from Sudan University of science and technology – Faculty of engineering in 1998, and a master degree in solid mechanics from Nile valley university (Atbara, Sudan) in 2003, and a PhD in structural engineering in 2017. He contributed in teaching some subjects in other universities such as Red Sea University (Port Sudan, Sudan), Kordofan University (Obayed, Sudan), Sudan University of Science and Technology (Khartoum, Sudan), Blue Nile University (Damazin, Sudan) and Kassala University (Kassala, Sudan). In addition, he supervised more than hundred and fifty under graduate studies in diploma and B.Sc. levels and about fifteen master theses. The author wrote about fifty engineering books written in Arabic language, and twenty five books written in English language and more than hundred research papers in fluid mechanics, thermodynamics, internal combustion engines and analysis of composite structures. He authored more than two thousands of lectures notes in the fields of mechanical, production and civil engineering He is currently an associated professor in Department of Mechanical Engineering, Faculty of Engineering and Technology, Nile Valley University Atbara, Sudan. His research interest and favorite subjects include structural mechanics, applied mechanics, control engineering and instrumentation, computer aided design, design of mechanical elements, fluid mechanics and dynamics, heat and mass transfer and hydraulic machinery. The author also works as a technical manager and superintendent of Al – Kamali mechanical and production workshops group which specializes in small, medium and large automotive overhaul maintenance and which situated in Atbara town in the north part of Sudan, River Nile State. E – mail address: osamakhayal66@nilevalley.edu.sd or osamamm64@gmail.com.

.

I want morebooks!

Buy your books fast and straightforward online - at one of world's fastest growing online book stores! Environmentally sound due to Print-on-Demand technologies.

Buy your books online at
www.morebooks.shop

Kaufen Sie Ihre Bücher schnell und unkompliziert online – auf einer der am schnellsten wachsenden Buchhandelsplattformen weltweit! Dank Print-On-Demand umwelt- und ressourcenschonend produziert.

Bücher schneller online kaufen
www.morebooks.shop

KS OmniScriptum Publishing
Brivibas gatve 197
LV-1039 Riga, Latvia
Telefax: +371 686 204 55

info@omniscriptum.com
www.omniscriptum.com

Printed by Books on Demand GmbH, Norderstedt / Germany